T0205550

Advances in Computer Vision and Pattern Recognition

Founding editor

Sameer Singh, Rail Vision, Castle Donington, UK

Series editor

Sing Bing Kang, Microsoft Research, Redmond, WA, USA

Advisory Board

Horst Bischof, Graz University of Technology, Austria
Richard Bowden, University of Surrey, Guildford, UK
Sven Dickinson, University of Toronto, ON, Canada
Jiaya Jia, The Chinese University of Hong Kong, Hong Kong
Kyoung Mu Lee, Seoul National University, South Korea
Yoichi Sato, The University of Tokyo, Japan
Bernt Schiele, Max Planck Institute for Computer Science, Saarbrücken, Germany
Stan Sclaroff, Boston University, MA, USA

More information about this series at http://www.springer.com/series/4205

Guy Gilboa

Nonlinear Eigenproblems in Image Processing and Computer Vision

 Springer

Guy Gilboa
Technion—Israel Institute of Technology
Haifa
Israel

ISSN 2191-6586   ISSN 2191-6594   (electronic)
Advances in Computer Vision and Pattern Recognition
ISBN 978-3-030-09339-6   ISBN 978-3-319-75847-3   (eBook)
https://doi.org/10.1007/978-3-319-75847-3

Printed on acid-free paper

This Springer imprint is published by the registered company Springer International Publishing AG
part of Springer Nature
The registered company address is: Gewerbestrasse 11, 6330 Cham, Switzerland

To my wife Mor.

Preface

What are Nonlinear Eigenproblems and Why are They Important?

A vast majority of real-world phenomena are nonlinear. Therefore, linear modeling serves essentially as a rough approximation and in some cases offers only shallow understanding of the underlying problem. This holds not only for modeling physical processes but also for algorithms, such as in image processing and computer vision, which attempt to process and understand the physical world, based on various 2D and 3D sensors.

Linear algorithms and transforms have reached tremendous achievements in signal and image processing. This was accomplished by very effective tools such as Fourier and Laplace transforms and based on the broad and deep mathematical foundations of linear algebra. Thus, very strong theories have been established over the years and efficient numerical methods have been developed.

Unfortunately, linear algorithms have their limitations. This is especially evident in signal and image processing, where standard linear methods such as Fourier analysis are rarely used in modern algorithms. A main reason is that images and many other signals have nonstationary statistics and include discontinuities (or edges) in the data. Therefore, standard assumptions of global statistics and smoothness of the data do not apply. The common practice in the image processing field is to develop nonlinear algorithms. These can roughly be divided into several main approaches:

- **Local linearization**. Applying a linear operator which is adaptive and changes spatially in the image domain. Examples: adaptive Wiener filtering [1, 2], bilateral filtering [3], and nonlocal means [4].
- **Hybrid linear–nonlinear**. Performing successive iterations of linear processing followed by simple nonlinear functions (such as thresholding, sign, and sigmoid functions). This branch includes many popular algorithms such as wavelet thresholding [5], dictionary and sparse representation approaches [6–8], and recently, deep convolutional neural networks [9, 10].
- **Spectral methods**. In this branch, one often constructs a data-driven graph and then performs linear processing using spectral graph theory [11], where the graph Laplacian is most commonly used. Examples: graph cuts [12], diffusion maps [13], and random-walker segmentation [14].
- **Convex modeling**. Algorithms based on convex optimization with non-quadratic functionals, such as total variation [15–17]. More details are given on this branch in this book.
- **Kernel-based**. Applying directly nonlinear kernel operators such as median, rank filers [18], and morphological filtering [19].

In this book, we take a fresh look at nonlinear processing through nonlinear eigenvalue analysis. This is still a somewhat unorthodox approach, since eigenvalue analysis is traditionally concerned with linear operators. We show how one-homogeneous convex functionals induce operators which are nonlinear and can be analyzed within an eigenvalue framework. The book has three essential parts. First, mathematical background is provided along with a summary of some classical variational algorithms for vision (Chaps. 2–3). The second part (Chaps. 4–7) focuses on the foundations and applications of the new multiscale representation based on nonlinear eigenproblems. In the last part (Chaps. 8–11), new numerical techniques for finding nonlinear eigenfunctions are discussed along with promising research directions beyond the convex case. These approaches may be valuable also for scientific computations and for better understanding of scientific problems beyond the scope of image processing.

In the following, we present in more details the intuition and motivation for formulating nonlinear transforms. We begin with the classical Fourier transform and its associated operator and energy. We ask how these concepts can be generalized in the nonlinear case? This can give the reader the flavor of topics discussed in this book and the approach taken to address them.

Basic Intuition and Examples

Fourier and the Dirichlet Energy

Let us first look at the classical case of the linear eigenvalue problem

$$Lu = \lambda u, \tag{1}$$

where L is a bounded linear operator and u is some function in a Hilbert space. For u admitting (1), we refer to as an eigenfunction, where λ is the corresponding eigenvalue.

In this book, we investigate nonlinear operators which can be formed based on regularizers. This can be done either by deriving the variational derivative of the regularizer (its subdifferential in the general convex nonsmooth case) or by using the proximal operator. Let us examine the standard quadratic regularizer, frequent in physics, and the basis of Tikhonov regularization, the Dirichlet energy

$$J_D = \frac{1}{2}\int |\nabla u|^2 dx. \tag{2}$$

The variational derivative is

$$\partial_u J_D = -\Delta u,$$

with Δ the Laplacian. The corresponding eigenvalue problem is

$$-\Delta u = \lambda u, \tag{3}$$

where Fourier basis yields the set of eigenfunctions (with appropriate boundary conditions).

Another classical result in this context is the relation to the Rayleigh quotient,

$$R(u) = \frac{\|\nabla u\|_{L^2}^2}{\|u\|_{L^2}^2}.$$

Taking the inner product with respect to u from both sides of (3), we get $\langle -\Delta u, u\rangle = \lambda\langle u, u\rangle$. Using integration by parts (or the divergence theorem), we have $\langle -\Delta u, u\rangle = \|\nabla u\|_{L^2}^2$ and thus for any eigenfunction u we obtain

$$\lambda = R(u).$$

Moreover, if we seek to minimize $R(u)$, with respect to u, we can formulate a constrained minimization problem, subject to $\|u\|_{L^2}^2 = 1$, and see (using Lagrange multipliers) that eigenfunctions are extremal functions of the Rayleigh quotient (hence, also its global minimizer is an eigenfunction).

From the above discussion, we see the connection between a quadratic regularizer (J_D) and the related linear operator ($-\Delta$). We notice that the corresponding eigenvalue problem induces a multiscale representation (Fourier) and that eigenfunctions are local minimizers of the Rayleigh quotient. The remarkable recent findings are that these relations generalize very nicely to the nonlinear case. We will now look at the total variation (TV) regularizer, which induces a nonlinear eigenvalue problem.

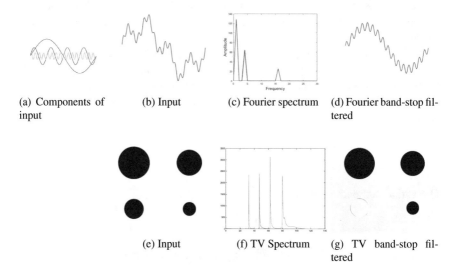

(a) Components of input (b) Input (c) Fourier spectrum (d) Fourier band-stop filtered

(e) Input (f) TV Spectrum (g) TV band-stop filtered

Fig. 1 Example of multiscale representation and filtering based on linear and nonlinear eigenfunctions. In the first row, three sine functions of different frequencies (left) are combined and serve as the input signal. A classical Fourier transform shows the spectrum containing three numerical delta functions. Filtering out the middle frequency (band-stop or notch filter) yields the result on the right. On the bottom row, we show a similar phenomenon based on the TV transform, where the input is a combination of four eigenfunctions in the shape of disks. Based on the TV transform, explained in Chap. 5, we get a spectrum with four numerical deltas, corresponding to each disk (from small to large radius). One can filter out the second smallest disk in an analog manner to Fourier, receiving the result on the right, preserving perfectly the three other disks, with some marginal errors due to grid discretization effects

The Total Variation Eigenvalue Problem

First, we would like to introduce the typical nonlinear eigenvalue problem associated with a bounded nonlinear operator T,

$$Tu = \lambda u. \tag{4}$$

This is a straightforward analog of (1) and will be investigated throughout this book, mainly in the context of convex optimization. Further generalizations to this equation will also be briefly addressed.

The L^1-type regularizer, which is analog to the Dirichlet energy, is the total variation functional defined by

$$J_{TV} = \int |\nabla u| dx. \tag{5}$$

Here, we use the simple (strong-sense) formulation. TV is very frequent in regularization of image processing problems, such as denoising, 3D reconstruction, stereo, and optical flow estimations. Chapter 3 gives more details on the properties of this functional. Based on the variational derivative of TV, $\partial_u J_{TV}$, we reach the following nonlinear eigenvalue problem:

$$-\text{div}\left(\frac{\nabla u}{|\nabla u|}\right) = \lambda u, \tag{6}$$

which is the analog of (3). For those less familiar with variational methods, we give the basic background in Sect. 1.3 on how the operator on the left side of (6) was derived. This operator is often referred to as the 1-Laplacian. We also note that this writing is somewhat informal (a subgradient inclusion is more precise, and the derivatives should be understood in the distributional sense).

Let u be an indicator function of a convex set C in $\mathbb{R}^2$. A fascinating result by Andreau et al. [20] is that u is an eigenfunction, in the sense of (6), if C admits the following simple geometric condition:

$$(\text{maximal curvature on } \partial C) \leq \frac{Per(C)}{|C|},$$

where ∂C is the boundary of C, $Per(C)$ is its perimeter and $|C|$ is its area. Checking this condition for the case of a disk of radius r (curvature $1/r$, $Per(C) = 2\pi r$, $|C| = \pi r^2$), we observe that any disk admits this condition. It essentially means that convex characteristics sets with smooth enough boundaries are TV eigenfunctions. Moreover, it was established that the eigenvalue is precisely the perimeter to area ratio

$$\lambda = \frac{Per(C)}{|C|}.$$

An alternative way to derive λ for any u is to follow the computation described earlier for the linear case. That is, we can take the inner product with respect to u of (6) and get

$$\lambda = \frac{J_{TV}(u)}{\|u\|_{L^2}^2}.$$

Indeed, for the specific case of a characteristic set, we obtain the equalities $Per(C) = J_{TV}(u)$ and $|C| = \|u\|_{L^2}^2$. The last equation is actually a generalized Rayleigh quotient, where one can show that eigenfunctions are extremal points, as in the linear case.

A natural question to ask is, can we also generalize a multiscale representation, based on TV eigenfunctions, in an analog manner to Fourier? If so, what are the properties and qualities of this representation? how well can it represent images? This book tries to address this fundamental question. A suggestion for a spectral representation suggested by the author and colleagues [21–23] is the spectral TV representation or the TV transform. A detailed explanation of the formalism is given in Chap. 5.

In the multiscale representation of the TV transform, scales (in the sense of eigenvalues) can be well separated, amplified, or attenuated, in a similar manner to classical Fourier filtering. In Fig. 1, we show a toy example of *TV band-stop filtering* along a comparison to standard Fourier filtering.

Graphs and Physics

Nonlinear eigenvalue problems related to TV are highly useful also for classification and learning, using graph data structures. The Cheeger constant is an isoperimetric value which essentially measures the degree of the dominant "bottleneck" in the graph. Perimeter of a set on a graph is directly related to the total variation on graphs. It was established that finding this bottleneck, or Cheeger cut, which is a NP-hard problem, can be well approximated by solving the 1-Laplacian eigenvalue problem

$$\Delta_1 u = \lambda \text{sign}(u),$$

where $\Delta_1 u$ is the 1-Laplacian on the graph. In Chap. 8, it will be explained how this can be used for segmentation and classification. Also, numerical algorithms for solving such problems will be discussed in Chap. 7. Finally, we will briefly mention the more general double-nonlinear eigenvalue problem

$$T(u) = \lambda Q(u),$$

where both T and Q can be nonlinear operators. In the case of $T = -\Delta$ (a linear operator) and nonlinear Q, there are several physical problems, such as soliton waves, which are modeled by solutions to these types of eigenvalue problems.

What is Covered in This Book?

This book first presents some of the basic mathematical notions, which are needed for later chapters. An effort is made to make the book self-contained so it is accessible to many disciplines. We then outline briefly the use of variational and flow-based methods to solve many image processing and computer vision algorithms (Chap. 3).

As total variation is an important functional, which is used throughout this book, we present its properties in more details (Chap. 4). We then define the concept of nonlinear eigenfunctions related to convex functionals and state some of the properties known today (an area still under active research, Chap. 5).

We proceed by going into a fundamental concept presented in this book of the spectral framework for one-homogeneous functionals. We show how eigenfunctions appear naturally in gradient descent and variational methods and that a spectral decomposition can be used for new representations of signals. The concept and motivation are discussed, as well as current theory on the subject. Applications of using this framework for denoising, texture processing, and image fusion are presented (Chaps. 6–7).

In the following chapter, we go deeper into the nonlinear eigenvalue problem and propose new ways to solve it using special flows which converge to eigenfunctions (Chap. 8).

We then go to graph-based and nonlocal methods, where a TV eigenvalue analysis gives rise to strong segmentation, clustering, and classification algorithms (Chap. 9).

Next, we present a new direction of how the nonlinear spectral concept can be generalized beyond the convex case, based on pixel decay analysis (Chap. 10). We are thus able to construct a spectral representation with different nonlinear denoisers and get different eigenmodes.

Relations to other image processing branches, such as wavelets and dictionary based, are discussed (Chap. 11). We conclude with the current open problems and outline future directions for the development of theory and applications related to nonlinear eigenvalue problems. In the appendix, we summarize some standard discretization and convex optimization methods, which are used to implement numerically such methods.

Haifa, Israel   Guy Gilboa

References

1. D.T. Kuan, A.A. Sawchuk, T.C. Strand, P. Chavel, Adaptive noise smoothing filter for images with signal-dependent noise. IEEE Trans. Pattern Anal. Mach. Intell. **7**(2), 165–177 (1985)
2. J.-S. Lee, Digital image enhancement and noise filtering by use of local statistics. IEEE Trans. Pattern Anal. Mach. Intell. **2**(2), 165–168 (1980)

3. C. Tomasi, R. Manduchi, Bilateral filtering for gray and color images, in *ICCV '98* (1998), pp. 839–846
4. A. Buades, B. Coll, J.-M. Morel, A review of image denoising algorithms, with a new one. SIAM Multiscale Model. Simul. **4**(2), 490–530 (2005)
5. D.L. Donoho, De-noising by soft-thresholding. IEEE Trans. Inf. Theory **41**(3), 613–627 (1995)
6. M. Elad, M. Aharon, Image denoising via sparse and redundant representations over learned dictionaries. IEEE Trans. Image Process, **15**(12), 3736–3745 (2006)
7. J. Mairal, F. Bach, J. Ponce, G. Sapiro, Online dictionary learning for sparse coding, in *Proceedings of the 26th Annual International Conference on Machine Learning* (ACM, 2009), pp. 689–696
8. M. Elad, *Sparse and Redundant Representations* (Springer, New York, 2010)
9. A. Krizhevsky, I. Sutskever, G.E. Hinton, Imagenet classification with deep convolutional neural networks, in *Advances in Neural Information Processing Systems* (2012), pp. 1097–1105
10. I.H. Witten, E. Frank, M.A. Hall, C.J. Pal, *Data Mining: Practical Machine Learning Tools and Techniques* (Morgan Kaufmann, Burlington, 2016)
11. F.R.K. Chung, in *Spectral Graph Theory*, vol. 92, American Mathematical Society (1997)
12. J. Shi, J. Malik, Normalized cuts and image segmentation. IEEE Trans. Pattern Anal. Mach. Intell. **22**(8), 888–905 (2000)
13. R.R. Coifman, S. Lafon, Diffusion maps. Appl. Comput. Harmon. Anal. **21**(1), 5–30 (2006)
14. L. Grady, Random walks for image segmentation. To appear in IEEE Trans. Pattern Anal. Mach. Intell. (2006)
15. G. Aubert, P. Kornprobst, *Mathematical Problems in Image Processing*, vol. 147 of Applied Mathematical Sciences, (Springer-Verlag, 2002)
16. L. Rudin, S. Osher, E. Fatemi, Nonlinear total variation based noise removal algorithms. Physica D, 60, 259–268 (1992)
17. A. Chambolle, V. Caselles, D. Cremers, M. Novaga, T. Pock, An introduction to total variation for image analysis. Theor. Found. Numer. Methods sparse recovery 9, 263–340 (2010)
18. G. Heygster, Rank filters in digital image processing. Comp. Graphics Image Process. **19**(2), 148–164 (1982)
19. L. Vincent, Morphological grayscale reconstruction in image analysis: Applications and efficient algorithms. IEEE Trans. Image Process. **2**(2), 176–201 (1993)
20. F. Andreu, C. Ballester, V. Caselles, J.M. Mazón, Minimizing total variation flow. Differ. Integral Equ. **14**(3), 321–360 (2001)
21. G. Gilboa, A spectral approach to total variation, in *SSVM 2013*, eds. by A. Kuijper et al., vol. 7893, Lecture Notes in Computer Science, (Springer, 2013), p. 36–47
22. G. Gilboa. A total variation spectral framework for scale and texture analysis. SIAM J. Imaging Sci. **7**(4), 1937–1961 (2014)
23. M. Burger, G. Gilboa, M. Moeller, L. Eckardt, D. Cremers, Spectral decompositions using one-homogeneous functionals. SIAM J. Imaging Sci. **9**(3), 1374–1408 (2016)

Acknowledgements

The author would like to thank the following people for their advice, help, and contributions, which made the writing of this book possible: Martin Burger, Michael Moeller, Jean-Francois Aujol, Martin Benning, Nicolas Papadakis, Tomer Michaeli, Ester Hait, Tal Feld, Shai Biton, Ido Cohen, Raz Nossek, Dikla Horesh, Oren Katzir, Alona Baruhov, and Nati Daniel.

The author acknowledges support by the Israel Science Foundation (Grant No. 718/15).

Contents

Chapter 1
Mathematical Preliminaries

1.1 Reminder of Very Basic Operators and Definitions

All definitions are given for the N-dimensional case. In images $N = 2$, in volumes (e.g., medical imaging) $N = 3$. Assume functions $u : X \subseteq \mathbb{R}^N \to \mathbb{R}$ and $\mathbf{v} : X \subseteq \mathbb{R}^N \to \mathbb{R}^N$, i.e., $\mathbf{v} = (v_1, \cdots, v_N)$, $v_i : X \subseteq \mathbb{R}^N \to \mathbb{R}$, $1 \le i \le N$.

Partial Derivative

$$\partial_{x_i} u := \lim_{h \to 0} \frac{u(x_1, .., x_i + h, .., x_N) - u(x_1, .., x_i, .., x_N)}{h}$$

Gradient

$$\nabla u := (\partial_{x_1} u, .., \partial_{x_N} u), \quad N\text{-dimensional vector.}$$

Directional Derivative Directional derivative of u at point x and direction ξ is defined as

$$\nabla_\xi u := \lim_{h \to 0} \frac{u(x + h\xi) - u(x)}{h}.$$

If u is differentiable at x, this directional derivative can be evaluated as

$$\nabla_\xi u = \nabla u \cdot \xi$$

Divergence

$$\text{div } \mathbf{v} := \sum_{i=1}^{N} \partial_{x_i} v_i(x)$$

© Springer International Publishing AG, part of Springer Nature 2018
G. Gilboa, *Nonlinear Eigenproblems in Image Processing and Computer Vision*, Advances in Computer Vision and Pattern Recognition,
https://doi.org/10.1007/978-3-319-75847-3_1

Laplacian

$$\Delta u := \mathrm{div}(\nabla u) = \sum_{i=1}^{N} \partial_{x_i}^2 u$$

where $\partial_{x_i}^2$ is a shortage of $\partial_{x_i} \partial_{x_i}$.

Structure Tensor

$$\mathscr{J}(u) := (\nabla u)(\nabla u)^T$$

Boundary Conditions (BC)

Very often image processes are described as a partial differential derivative (the most known is anisotropic diffusion). In these and other cases, we ought to set the boundary conditions. The main two types are defined below.
Assume $u(x) : \Omega \subset \mathbb{R}^2 \to R$ is the image and $\partial\Omega$ is the boundary of Ω, then:

Dirichlet BC:

$$u(x) = F(x), \ \forall x \in \partial\Omega.$$

Neumann BC:

$$\nabla_{\mathbf{n}} u(x) = F(x), \ \forall x \in \partial\Omega,$$

where $\mathbf{n}$ is the direction normal to the boundary $\partial\Omega$. Usually, in image processing, one uses zero Neumann BC: $\nabla_{\mathbf{n}} u(x) = 0$.

1.1.1   Integration by Parts (Reminder)

To derive it, we can begin with the identity: $(uv)' \equiv u'v + uv'$, where $'$ denotes spatial differentiation. For 1D, integration of both sides gives $\int_A^B \frac{d}{dx}(uv)dx = \int_A^B u'v dx + \int_A^B uv' dx$, and therefore

$$\int_A^B u'v dx = (uv)|_A^B - \int_A^B uv' dx.$$

In any dimension, for a function $u(x)$ and a vector function $\mathbf{v} = (v_1(x), \ldots, v_N(x))$ we have:

$$\int_\Omega \nabla u \cdot \mathbf{v} dx = \int_{\partial\Omega} u(\mathbf{v} \cdot \mathbf{n})ds - \int_\Omega u \ \mathrm{div} \ \mathbf{v} dx,$$

where $\mathbf{n}$ is the outward unit normal vector to $\partial\Omega$. For $(\mathbf{v} \cdot \mathbf{n})|_{\partial\Omega} = 0$ we get

$$\int_\Omega \nabla u \cdot \mathbf{v} dx = - \int_\Omega u \ \mathrm{div} \ \mathbf{v} dx.$$

1.1.2 Distributions (Reminder)

Test Function:

Let φ be a test function with the following properties:

1. φ is smooth (infinitely differentiable in principle).
2. φ has a compact support (is identically zero outside some bounded interval).

The set of all test functions is denoted as $\mathscr{C}_c^\infty(\Omega)$ where $\mathscr{C}$ is for continuous, the subscript $_c$ is for compact support (which is Ω) and the superscript $^\infty$ stands for the existence of kth order derivatives for all natural $k > 0$.

Distributional Derivatives: This is a generalization of integrals of functions based on derivatives, where the derivatives are not defined everywhere. In cases where we have an integral of an expression with derivatives, we can write it using a test function and use integration by parts to "move" the derivative to the test function. We thus have an alternative way to define derivatives in a weak sense which holds also in cases where standard derivative notions are undefined.

The distributional (or weak) derivative of order k of f is g if for all $\varphi \in \mathscr{C}_c^\infty(\Omega)$ holds

$$\int_\Omega g(x)\varphi(x)dx = (-1)^k \int_\Omega f(x)\partial_x^k \varphi(x)dx.$$

Example Consider the function $f(x) \in \mathbb{R}$ defined as $f = 0$ if $x \leq 0$ and $f = x$ if $x > 0$. Its first-order distributional derivative can be computed using

$$-\int_\mathbb{R} f(x)\varphi'(x)dx = -\int_0^\infty x\varphi'(x)dx = \int_0^\infty \varphi(x)dx = \int_\mathbb{R} H(x)\varphi(x)dx,$$

where $H(x)$ is the Heaviside function $H(x) = 0$ if $x \leq 0$ and $H(x) = 1$ if $x > 0$. Thus, $H(x)$ is the weak derivative of $f(x)$.

1.2 Some Standard Spaces

In this section, F is set of functions from $\mathbb{R}^N$ to $\mathbb{R}$ and V is a linear space over the field $\mathbb{R}$ or $\mathscr{C}$ (in this book we deal only with real-valued functions and spaces).

Metric Space:

A metric space is an ordered pair (F, d) where d is a positive function from $F \times F$ to $\mathbb{R}$ that holds

1. $d(x, y) \geq 0$
2. $d(x, y) = 0 \Leftrightarrow x = y$
3. $d(x, y) = d(y, x)$
4. $d(x, y) \leq d(x, z) + d(z, y)$

for all $x, y, z \in F$. d is **semimetric** if it holds only the first three properties above (there are some weaker definition to the fourth property but they are not necessary in the scope of this book).

Norm Space:

(V, p) where $p : V \Rightarrow \mathbb{R}$ is a norm space if p holds

1. $p(av) = |a| p(v)$
2. $p(v + u) \leq p(v) + p(u)$
3. $p(v) \leq 0$
4. if $p(v) = 0$ then $v = 0$

$\forall a \in \mathbb{R}$ and $\forall v, u \in V$. p is **seminorm** if it holds only the first three properties above.

Every norm space can induce metric space by the definition $d(x, y) = p(v - u)$.

Banach Space

Norm space is a Banach space if the induced metric space is complete.

Inner Product Space:

Inner product space is a vector space V with inner product $< \cdot, \cdot >$ mapping pairs of elements from V to a real or complex number, where the following properties hold:

1. $< x, y > = < \bar{y}, x >$.
2. $< ax, y > = a < x, y >$.
3. $< x + y, z > = < x, z > + < y, z >$.
4. $< x, x > \geq 0$.
5. if $< x, x > = 0$ then $x = 0$.

$\forall a \in \mathbb{R}$ and $\forall x, y, z \in V$ where $\bar{w}$ is the complex conjugate of w.

Every inner product space can induce a norm space by the definition $p(x) = \sqrt{< x, x >}$ and therefore a metric space by $d(x, y) = \sqrt{p(x - y)}$.

Hilbert Space

Inner product space is a Hilbert space if the metric space induced from the inner product is complete.

1.3   Euler–Lagrange

Functional:

Let U be a Banach space, $u, v \in U$ and F is a mapping from U to $\mathbb{R}$.

Gâteaux derivative Gâteaux derivative is a derivative of a functional and in some sense can be seen as an extension of functions' directional derivative. If $v, u \in U$ the Gâteaux derivative is defined as

$$F'(u) = \lim_{\lambda \to 0^+} \frac{F(u + \lambda v) - F(u)}{\lambda}.$$

In the variational context, we refer to v as the variation and assume zero contribution on the boundary, $v|_{\partial\Omega} = 0$. We also assume that F is Gâteaux differentiable, that is for all v we reach the same limit.

The equation

$$F'(u) = 0 \tag{1.7}$$

is called the Euler–Lagrange (E–L) equation. We usually refer to $F'(u)$ as the E–L of functional $F(u)$.

If we have a minimization problem $\inf_{u \in X} F(u)$ (where inf is the infimum), the solution u admits Eq. (1.7). On the other hand, if F is convex, then a solution u of Eq. (1.7) is also a solution of the minimization problem. For more details see [1], Chap. 2, p. 37.

1.3.1 E–L of Some Functionals

Let a functional F be defined by

$$F(u) = \int_{\Omega} f(x, u(x), \nabla u(x)) dx.$$

We view f as a function of three terms: $f(x, u, \xi)$ in $\mathbb{R}^N$, where $x = (x_1, .., x_N)$, $\xi = (\xi_1, .., \xi_N)$. The derivative of the functional (E–L) is

$$F'(u) = \partial_u F(u) = \frac{\partial f(x, u, \nabla u)}{\partial u} - \sum_{i=1}^{N} \frac{\partial}{\partial x_i} \left(\frac{\partial f(x, u, \nabla u)}{\partial \xi_i} \right) \tag{1.8}$$

For precise smoothness and growth conditions, see [1] (same place as above).

For regularization, we are usually concerned only with derivatives of u, thus we can have a simplified formulation. Let $J(u)$ be a regularization functional of the form:

$$J(u) = \int_{\Omega} \Phi(|\nabla u(x)|) dx. \tag{1.9}$$

Then the E–L, or the gradient, of the functional is

$$\partial_u J(u) = -\operatorname{div}\left(\frac{\Phi'(|\nabla u|)}{|\nabla u|}\nabla u\right).\qquad(1.10)$$

To minimize a convex functional, we can perform a gradient descent, initializing with some $f(x)$. As a continuous time process, the gradient descent evolution is

$$u_t = -\partial_u J(u) = \operatorname{div}\left(\frac{\Phi'(|\nabla u|)}{|\nabla u|}\nabla u\right),\quad u|_{t=0} = f(x).\qquad(1.11)$$

This is exactly a nonlinear, Perona–Malik style, diffusion equation, where the diffusion coefficient c is

$$c(|\nabla u|) = \frac{\Phi'(|\nabla u|)}{|\nabla u|}.$$

Thus, we can view nonlinear diffusion as a process which minimizes a regularizing functional.

Remark Note that the original P–M diffusion coefficients can be shown as minimizing non-convex energy functionals, thus their descent can theoretically get stuck in a local minimum. However it was shown that discretization or the Catte et al. gradient smoothing avoids this problem.

1.3.2 Some Useful Examples

Linear diffusion: $\Phi(s) = \frac{1}{2}s^2$.

$$J_{Lin}(u) = \frac{1}{2}\int_{\Omega}|\nabla u(x)|^2 dx.$$

The gradient descent is the linear diffusion equation $c = 1$:

$$u_t = \operatorname{div}(\nabla u) = \Delta u.$$

TV Flow: $\Phi(s) = s$.

$$J_{TV}(u) = \int_{\Omega}|\nabla u(x)|dx.$$

The gradient descent is called the TV-flow [2, 3], which has special properties, qualitatively similar to the ROF model [4]. We can view it as nonlinear diffusion with $c = \frac{1}{|\nabla u|}$ (note the unbounded value at zero gradient):

$$u_t = \operatorname{div}\left(\frac{\nabla u}{|\nabla u|}\right).$$

Perona–Malik: For the diffusion coefficient $c(s) = \frac{1}{1+(s/k)^2}$ we get $\Phi(s) = \frac{1}{2}k^2 \log(1 + (s/k)^2)$.

$$J_{PM}(u) = \frac{1}{2}k^2 \int_\Omega \log(1 + (|\nabla u(x)|/k)^2)dx.$$

This is a non-convex functional. The gradient descent is naturally the P–M equation:

$$u_t = \mathrm{div}\left(\frac{\nabla u}{1 + (|\nabla u|/k)^2}\right).$$

1.3.3 E–L of Common Fidelity Terms

L^2 norm square: $F_{(L^2)^2}(u) = \|u\|_{L^2}^2 = \int_\Omega (u(x))^2 dx.$

$$\partial_u F_{(L^2)^2}(u) = 2u.$$

L^{1}norm: $F_{L^1}(u) = \|u\|_{L^1} = \int_\Omega |u(x)|dx.$

$$\partial_u F_{L^1}(u) = \mathrm{sign}(u).$$

where sign is the signum function: $\mathrm{sign}(q) = 1$ for $q > 0$, $= -1$ for $q < 0$. For $q = 0$, we often keep the vague definition $\mathrm{sign}(0) \in [-1, 1]$ to keep it consistent with the notion of subdifferential.

1.3.4 Norms Without Derivatives

Very useful norms are the L^p norms, defined as

$$\|u\|_{L^p} \equiv \left(\int_\Omega (u(x))^p dx\right)^{1/p}.$$

Interesting special cases are

1. **L^2**

$$\|u\|_{L^2} \equiv \sqrt{\int_\Omega (u(x))^2 dx}.$$

2. $\mathbf{L^1}$

$$\|u\|_{L^1} \equiv \int_\Omega |u(x)|dx.$$

3. $\mathbf{L^\infty}$

$$\|u\|_{L^\infty} \equiv \max_{x \in \Omega} |u(x)|.$$

4. $\mathbf{L^0}$

$$\|u\|_{L^0} \equiv \int_\Omega I(u(x))dx,$$

where $I(q) = 1$ if $q \neq 0$ and 0 if $q = 0$. Note that L^0 is not an actual norm, as it does not admit the triangle inequality.

The Frobenius for matrices is essentially the 2-norm of the matrix elements. For a real-valued matrix A it can be written compactly as

$$\|A\|_F = \sqrt{Tr(AA^T)},$$

where A^T is the transpose of the matrix (or conjugate transpose for complex-valued elements).

The norms above are usually used as *fidelity terms*, penalizing distance to the input image.

Cauchy–Schwarz inequality: We define the L^2 inner product as

$$\langle u, v \rangle = \int_\Omega u(x)v(x)dx,$$

then we have the following inequality:

$$|\langle u, v \rangle| \leq \|u\|_{L^2} \|v\|_{L^2}.$$

Holder inequality: The more general case, for two norms L^p, L^q where $\frac{1}{p} + \frac{1}{q} = 1$, $p, q \in [1, \infty]$ we have

$$|\langle u, v \rangle| \leq \|u\|_{L^p} \|v\|_{L^q}.$$

1.3.5   Seminorms with Derivatives

We assume a smooth signal (where the derivatives are well defined).

1. **H^1**

$$|u|_{H^1} \equiv \sqrt{\int_\Omega (\nabla u(x))^2 dx}.$$

2. **TV**

$$|u|_{TV} \equiv \int_\Omega |\nabla u(x)| dx.$$

These seminorms are usually used as *smoothness terms*, penalizing non-smoothness of the signal.

In order to use them as norms, the corresponding L^p norm is taken: $H^1 + L^2$ or $TV + L^1$.

We will later see that weaker smoothness conditions are sufficient, so TV can be well defined also on discontinuous functions (using distributions to define weak derivatives).

1.4 Convex Functionals

We will start by defining several basic notions in convex analysis. They serve as very good mathematical tools to better understand minimization of convex energies. Some numerical algorithms rely on this theory, like Chambolle's projection algorithm [5]. This is a sketchy outline on the topic. More can be found in the following books [6–8].

Convex set: A set $C \subseteq \Omega$ is convex iff (if and only if) for any x, $y \in C$ and $\alpha \in [0, 1]$

$$\alpha x + (1 - \alpha)y \in C.$$

1.4.1 Convex Function and Functional

A function $f(x)$ is convex iff for any x, $y \in \Omega$ and $\alpha \in [0, 1]$

$$f(\alpha x + (1 - \alpha)y) \leq \alpha f(x) + (1 - \alpha)f(y).$$

We can define the *epigraph* of the function as the set of points above or equal to f:

$$\text{epi}(f) = \{(x, w) : x \in \Omega, \, w \in \mathbb{R}, \, w \geq f(x)\}.$$

Then an equivalent definition for convexity: f is convex iff epi(f) is a convex set.

Functionals can be defined in the same way. A functional $J(u(x))$ is convex iff for any $u_1(x), u_2(x)$, $x \in \Omega$ and $\alpha \in [0, 1]$

$$J(\alpha u_1(x) + (1 - \alpha)u_2(x)) \leq \alpha J(u_1(x)) + (1 - \alpha)J(u_2(x)).$$

It is easy to show that functionals of the form $J = \int_\Omega f(u(x))dx$ are convex if $f(\cdot)$ is convex.

As q^2, $|q|$ and $\sqrt{q^2 + \varepsilon^2}$ are all convex in q, we can assign $q = |\nabla u(x)|$ and see that

- $J_{H^1} = \int_\Omega |\nabla u|^2 dx$
- $J_{TV} = \int_\Omega |\nabla u| dx$
- $J_{TV-\varepsilon} = \int_\Omega \sqrt{|\nabla u|^2 + \varepsilon^2}dx$

are all convex functionals.

Remark 1 We say that a function is strictly convex if for $\forall x \neq y, \alpha \in (0, 1)$ the inequality is strict:

$$f(\alpha x + (1 - \alpha)y) < \alpha f(x) + (1 - \alpha)f(y).$$

The same is for functionals. Therefore J_{H^1} is strictly convex, whereas J_{TV} is not, which yields some unique properties.

Remark 2 If f is twice differentiable in 1D then iff $f'' \geq 0$ then f is convex. If $f'' > 0$ then f is strictly convex (but not vice versa, see e.g., $f(x) = x^4$ at $x = 0$). For N dimensions, if the $N \times N$ Hessian matrix $H = (\partial_{x_i x_j} f)$, $(i, j = 1, .., N)$ is positive semi-definite then the function is convex.

1.4.2 Why Convex Functions Are Good?

1. All local minima are global minima.
2. Very well understood functions.
3. There are many efficient numerical methods to minimize convex functionals.
4. Strong duality tools are available both for efficient numerics and for developing theory.

1.4.3 Subdifferential

Extension of the derivative to non-differentiable cases. For functions we define the following set:

$$\partial f(x) := \{m : f(y) \geq f(x) + m(y - x), \forall y \in \Omega\}. \qquad (1.12)$$

- Every element m in the set is a called a *subgradient* and the entire set is called the *subdifferential*.

- If f is differentiable (at least once) then $\partial f(x)$ has exactly one element which is the gradient at the point $\partial f(x) = \nabla f(x)$.
- A point x_0 is a global minimum of f iff zero is contained in the subdifferential at that point $0 \in \partial f(x_0)$.

For functionals, we assume u, v are in some space X and p is in the dual space X^* and define

$$\partial J(u) := \left\{ p \in X^* : J(v) \geq J(u) + \langle p, v - u \rangle, \forall v \in X \right\}. \tag{1.13}$$

1.4.4 Duality—Legendre–Fenchel Transform

For functions:

$$f^*(m) := \sup_{x \in \Omega} \{mx - f(x)\}. \tag{1.14}$$

For functionals:

$$J^*(p) := \sup_{u \in X} \{\langle p, u \rangle - J(u)\}. \tag{1.15}$$

Some properties:

1. J^* is convex.
2. $J^{**} = J$.
3. For differentiable J we have $\partial_u J(u) = p$, $\partial_p J^*(p) = u$.
4. if J is not convex, J^* and J^{**} are still convex, J^{**} is the largest convex function satisfying $J^{**}(u) \leq J(u)$ and is called the convex envelope of $J(u)$.

1.5 One-Homogeneous Functionals

Below we present some essential properties of convex one-homogeneous functionals. These functionals include all norms and seminorms, including regularizers like total variation and total generalized variation (TGV) which are discussed and analyzed throughout this book.

1.5.1 Definition and Basic Properties

The assumption that $J : \mathbb{R}^n \to \mathbb{R}$ is absolutely one-homogeneous

$$J(su) = |s| J(u) \quad \forall s \in \mathbb{R}, \forall u \in \mathscr{X}. \tag{1.16}$$

Moreover, we shall assume that J is convex throughout the whole book.

Taken from [9]. Under our above assumptions, we can interpret J as a seminorm, respectively a norm on an appropriate subspace.

Lemma 1.1   *A functional J as above is a seminorm and its nullspace*

$$\mathcal{N}(J) = \{u \in \mathbb{R}^n \mid J(u) = 0\}$$

is a linear subspace. In particular there exist constants $0 < c_0 \leq C_0$ such that

$$c_0 \|u\| \leq J(u) \leq C_0 \|u\|, \qquad \forall u \in \mathcal{N}(J)^\perp. \tag{1.17}$$

Proof First of all we observe that J is nonnegative and absolutely one-homogeneous due to our above definitions, so that it suffices to verify the triangle inequality. From the convexity and absolute one-homogeneity, we have for all $u, v \in \mathcal{X}$

$$J(u+v) = 2J\left(\frac{1}{2}u + \frac{1}{2}v\right) \leq 2\left(\frac{1}{2}J(u) + \frac{1}{2}J(v)\right) = J(u) + J(v).$$

The fact that the nullspace is a linear subspace is a direct consequence, and the estimate (1.17) follows from the norm equivalence in finite-dimensional space.

Lemma 1.2   *Let J be as above, then for each $u \in \mathbb{R}^n$ and $v \in \mathcal{N}(J)$ the identity*

$$J(u+v) = J(u) \tag{1.18}$$

holds.

Proof Using the triangle inequality we find

$$J(u+v) \leq J(u) + J(v) = J(u),$$
$$J(u) = J(u+v-v) \leq J(u+v) + J(-v) = J(u+v),$$

which yields the assertion.

We continue with some properties of subgradients:

Lemma 1.3   *Let $u \in \mathbb{R}^n$, then $p \in \partial J(u)$ if and only if*

$$J^*(p) = 0 \quad and \quad \langle p, u \rangle = J(u). \tag{1.19}$$

A common reformulation of Lemma 1.3 is the characterization of the subdifferential of an absolutely one-homogeneous J as

$$\partial J(u) = \{p \in \mathcal{X}^* \mid J(v) \geq \langle p, v \rangle \ \forall v \in \mathcal{X}, \ J(u) = \langle p, u \rangle\}. \tag{1.20}$$

Remark 1.1 A simple consequence of the characterization of the subdifferential of absolutely one-homogeneous functionals is that any $p \in \partial J(0)$ with $p \notin \partial J(u)$ meets $J(u) - \langle p, u \rangle > 0$.

For u, p in ℓ^2, the Cauchy–Schwarz inequality and (1.19) directly implies

$$J(u) \leq \|u\| \|p(u)\|, \quad \forall p(u) \in \partial J(u). \tag{1.21}$$

Additionally, we can state a property of subgradients relative to the nullspace of J with a straightforward proof:

Lemma 1.4 *Let $p \in \partial J(0)$ and $J(u) = 0$, then $\langle p, u \rangle = 0$. Consequently $\partial J(u) \subset \partial J(0) \subset \mathcal{N}(J)^{\perp}$ for all $u \in \mathbb{R}^n$.*

The nullspace of J and its orthogonal complement will be of further importance in the sequel of the book. In the following, we will denote the projection operator onto $\mathcal{N}(J)$ by P_0 and to $\mathcal{N}(J)^{\perp}$ by $Q_0 = Id - P_0$. Note that as a consequence of Lemma 1.2, we have $J(u) = J(Q_0 u)$ for all $u \in \mathbb{R}^n$.

Lemma 1.5 *For absolutely one-homogeneous J, the following identity holds*

$$\bigcup_{u \in \mathbb{R}^n} \partial J(u) = \partial J(0) = \{p \in \mathbb{R}^n \mid J^*(p) = 0\}. \tag{1.22}$$

Moreover $\partial J(0)$ has nonempty relative interior in $\mathcal{N}(J)^{\perp}$ and for any p in the relative interior of $\partial J(0)$, we have $p \in \partial J(u)$ if and only if $J(u) = 0$.

Proof We have $p \in \partial J(0)$ if and only if $\langle p, u \rangle \leq J(u)$ for all u. Since equality holds for $u = 0$, this is obviously equivalent to $J^*(p) = 0$. Since we know from Lemma 1.3 that $\partial J(u)$ is contained in $\{p \in \mathbb{R}^n \mid J^*(p) = 0\}$ and the union also includes $u = 0$ we obtain the first identity. Let $p \in \mathcal{N}(J)^{\perp}$ with $\|p\| < c_0$ sufficiently small. Then we know by the Cauchy–Schwarz inequality and (1.17) that

$$\langle p, u \rangle = \langle p, Q_0 u \rangle \leq \|p\| \|Q_0 u\| \leq \frac{\|p\|}{c_0} J(Q_0 u) < J(Q_0 u) = J(u)$$

for all u with $Q_0 u \neq 0$. Finally, let p be in the relative interior of $\partial J(0)$ and $p \in \partial J(u)$. Since there exists a constant $c > 1$ such that $cp \in \partial J(0)$, we find $c \langle p, u \rangle = c J(u) \leq J(u)$ and consequently $J(u) = 0$.

Conclusion 1 *Using (1.17) as well as Lemma 1.4, we can conclude that for any $p \in \partial J(0)$ we have*

$$\|p\|^2 \leq J(p) \leq C_0 \|p\|,$$

such that $\|p\| \leq C_0$ holds for all possible subgradients p.

As usual, e.g., in the Fenchel–Young Inequality, one can directly relate the characterization of the subdifferential to the convex conjugate of the functional J.

Lemma 1.6 *The convex conjugate of an absolutely one-homogeneous functional J is the characteristic function of the convex set $\partial J(0)$.*

Proof Due to (1.20), we have

$$\partial J(0) = \{p \in \mathscr{X}^* \mid J(v) - \langle p, v \rangle \geq 0 \ \forall v \in \mathscr{X}\}.$$

The definition
$$J^*(p) = \sup_u (\langle p, u \rangle - J(u))$$

tells us that if $p \in \partial J(0)$ the above supremum is less or equal to zero and the choice $u = 0$ shows that $J^*(p) = 0$. If $p \notin \partial J(0)$ then there exists a u such that $\langle p, u \rangle - J(u) > 0$ and the fact that

$$\langle p, \alpha u \rangle - J(\alpha u) = \alpha(\langle p, u \rangle - J(u))$$

holds for $\alpha \geq 0$ yields $J^*(p) = \infty$.

References

1. G. Aubert, P. Kornprobst, *Mathematical Problems in Image Processing*, vol. 147, Applied Mathematical Sciences (Springer, Berlin, 2002)
2. F. Andreu, C. Ballester, V. Caselles, J.M. Mazón, Minimizing total variation flow. Differ. Integral Equ. **14**(3), 321–360 (2001)
3. G. Bellettini, V. Caselles, M. Novaga, The total variation flow in R^N. J. Differ. Equ. **184**(2), 475–525 (2002)
4. L. Rudin, S. Osher, E. Fatemi, Nonlinear total variation based noise removal algorithms. Phys. D **60**, 259–268 (1992)
5. A. Chambolle, An algorithm for total variation minimization and applications. JMIV **20**, 89–97 (2004)
6. R.T. Rockafellar, *Convex Analysis* (Princeton university press, Princeton, 1997)
7. I. Ekeland, R. Temam, *Convex Analysis and Variational Problems* (Elsevier, Amsterdam, 1976)
8. D.P. Bertsekas, A. Nedić, A.E. Ozdaglar, *Convex Analysis and Optimization* (Athena Scientific, Belmont, 2003)
9. Martin Burger, Guy Gilboa, Michael Moeller, Lina Eckardt, Daniel Cremers, Spectral decompositions using one-homogeneous functionals. SIAM J. Imaging Sci. **9**(3), 1374–1408 (2016)

Chapter 2
Variational Methods in Image Processing

2.1 Variation Modeling by Regularizing Functionals

Data measurement is an inaccurate process. The sensors are of finite accuracy, there is additive photonic and electrical noise. There can be some gain difference between the sensing elements (e.g., camera pixels). There are also specific artifacts related to the technology. For instance, in computed tomography (CT), a medical imaging scanner, there are metal artifacts (metal absorbs most of the X-ray energy, affecting the tomographic computations) and beam-hardening artifacts (the X-ray spectrum changes along its path through the body, changing the attenuation coefficient of the tissues). For most sensing modalities, there are blur artifacts due to inaccurate focus, movement of the sensors or of the objects and so on. In some cases, there are missing data due to occlusions or limited capabilities of the acquisition technology.

In all the above scenarios, one needs to use some prior information regarding the observed objects and acquisition process in order to obtain better estimations from the raw data. A canonical way to formulate this is to have a cost function which penalizes deviation from the data, E_{data}, and a cost function which penalizes deviation from the model, E_{model}. One then tries to optimize (find the minimizer) of the combined energy

$$E_{Total} = E_{Model}(u) + \alpha E_{Data}(u, f),$$

where $\alpha > 0$ is a scalar weight between the two terms (α increases as one assumes the measured data is more reliable).

In the case of noisy measurements, one can often relate the model to some definition of smoothness of the solution u. The data term is formulated as a distance or fidelity to the measured input:

$$E_{Total} = E_{Smoothness}(u) + \alpha E_{Fidelity}(u, f),$$

G. Gilboa, *Nonlinear Eigenproblems in Image Processing and Computer Vision*, Advances in Computer Vision and Pattern Recognition, https://doi.org/10.1007/978-3-319-75847-3_2

where f is the input image and u is the variable image for which the energy is minimized. The smoothness term involves derivatives and the fidelity term is often an L^p norm.

Tikhonov regularization: One of the pioneers of this approach was Tikhonov [1] who formulated the following classical regularization problem:

$$E_{Tik} = |u|^2_{H^1} + \alpha \| f - u \|^2_{L^2}.$$

In this case, smoothness is defined in a very intuitive way of penalizing high derivatives. The fidelity term is based on the Euclidean norm, which is the classical notion of distance. Squaring the L^2 norm is mainly for computational purposes, this yields a very simple pointwise optimality condition. The Tikhonov method works very well for data which is smooth in nature (such as wave signals). However many signals, especially in imaging, contain edges and sharp transitions. Their nature can be piecewise constant or piecewise smooth. A regularization based on H^1 will penalize severely such edges (as they contain very high derivatives) and the resulting solution will be blurry. For this reason, Rudin, Osher, and Fatemi introduced the seminal ROF regularization.

Total variation regularization [2]:

$$E_{TV} = |u|_{TV} + \alpha \| f - u \|^2_{L^2}.$$

This form of smoothness term, based on the TV seminorm, is more appropriate for regularizing images and signals with edges. This term is still very high for noisy signals (and therefore tends to reduce noise). But now signals with edges are not penalized much and the optimal solution tends to be piecewise smooth. Regarding the fidelity term, it was kept as in the Tikhonov formulation. It essentially assumes the noisy measurement is relatively close to the real data, and there are no outliers. This is true for uniform or white Gaussian noise. If an image has outliers (completely black or white pixels for example), a L^2 fidelity penalizes severely large deviations from them, and such artifacts will not be removed well.

Total variation deconvolution: A kernel H is assumed to convolve the input image: $f = g * H + n$, where g is the clean image, n is noise. The fidelity term changes to have the following energy:

$$E_{TV-deconv} = |u|_{TV} + \alpha \| f - u * H \|^2_{L^2}.$$

This change in the fidelity term means that we are now measuring the distance between the blurred solution and the input. This method yields a very stable and robust way of performing the ill-posed problem of deconvolution, without getting into strong noise amplification artifacts which are common in standard inverse-filtering methods.

TV-L1 [3, 4]: To solve the problem of removing outliers, a change of the fidelity term to the L^1 norm was proposed by Nikolova:

$$E_{TV-L1} = |u|_{TV} + \alpha \|f - u\|_{L^1}.$$

It was shown that such a regularization is contrast invariant. That is, it penalizes fine scales and keeps large scales, regardless of their local contrast. Thus, small outlier pixels can be removed well, as well as standard noise. In this case, however, one should take into account that fine contrasted details, which are part of the clean signal, will also be smoothed out.

2.1.1 Regularization Energies and Their Respective E-L

It is now straightforward to obtain the E-L of some standard regularization models, based on Sect. 1.3. We will write the negative expression $-\partial_u E$ which is useful for the gradient descent equation.

2.1.1.1 TV Denoising

$$E_{TV} = |u|_{TV} + \alpha \|f - u\|_{L^2}^2.$$

$$-\partial_u E_{TV} = \mathrm{div}\left(\frac{\nabla u}{|\nabla u|}\right) + 2\alpha(f - u).$$

Epsilon approximation: As for $\nabla u = 0$ the $E - L$ expression, in the strong-sense formulation, is undefined, for numerical purposes an epsilon approximation is often used:

$$E_{TV-\varepsilon} = |u|_{TV-\varepsilon} + \alpha \|f - u\|_{L^2}^2,$$

where

$$|u|_{TV-\varepsilon} = \int_\Omega \sqrt{|\nabla u(x)|^2 + \varepsilon^2}\, dx.$$

The respective E-L is

$$-\partial_u E_{TV-\varepsilon} = \mathrm{div}\left(\frac{\nabla u}{\sqrt{|\nabla u(x)|^2 + \varepsilon^2}}\right) + 2\alpha(f - u).$$

Note that a more accurate formulation is to use a weak-sense expression and subgradients. Numerical methods based on dual formulations do not require an epsilon approximation. So as $|\nabla u| \gg \varepsilon$, we approximate well the total variation descent. Near zero gradient, for $|\nabla u| \ll \varepsilon$ we approach linear diffusion with $c \to \frac{1}{\varepsilon}$.

Numerical implementation problem: We recall that for explicit scheme the time step is bounded by the CFL condition, which is inversely proportional to the diffusion coefficient. Therefore, the time step Δt is proportional to ε. Thus as we model more precisely TV, we need to use much smaller time steps, or many more numerical iterations to simulate the same evolution time.

Alternative semi-implicit schemes for TV minimization can be much more efficient. Such methods are shown in the Appendix.

2.1.1.2 TV Deconvolution

As we saw earlier, in the case of deconvolution a kernel $H(x)$ is assumed to convolve the input image. The fidelity term changes and we obtain the following energy:

$$E_{TV-deconv} = |u|_{TV} + \alpha \|f - u * H\|_{L^2}^2.$$

$$-\partial_u E_{TV-deconv} = \text{div}\left(\frac{\nabla u(x)}{|\nabla u(x)|}\right) + 2\alpha H(-x) * (f(x) - H(x) * u(x)).$$

Here also, the TV term can be replaced by its epsilon approximation.

2.1.1.3 TV-L1 Outlier Removal

$$E_{TV-L1} = |u|_{TV} + \alpha \|f - u\|_{L^1}.$$

$$-\partial_u E_{TV-L1} = \text{div}\left(\frac{\nabla u}{|\nabla u|}\right) + \alpha \, \text{sign}(f - u).$$

In this case, both the TV term and the sign term are epsilon approximated.

2.2 Nonlinear PDEs

2.2.1 *Gaussian Scale Space*

The scale space approach was suggested as a multi-resolution technique for image structure analysis.

2.2.1.1 Axiomatic Approach

For low-level vision processing, certain requirements were set in order to construct an uncommitted front-end [5]:

- Linearity (no previous model)
- Spatial shift invariance
- Isotropy
- Scale invariance (no preferred size)

2.2.1.2 Linear Diffusion

The unique operator admitting all these requirements is a convolution with a Gaussian kernel. In order to be scale invariant, all scales were to be considered. Therefore the Gaussian convolution was to be applied to the input at all scales (with standard deviation of Gaussian kernel ranging from 0 to ∞). The diffusion process (which is also called the heat equation) is equivalent to a smoothing process with a Gaussian kernel. In this context, the linear diffusion equation was used:

$$u_t = c\Delta u, \quad u|_{t=0} = f, \quad c > 0 \in \mathbb{R}. \tag{2.1}$$

This introduced a natural continuous scale dimension t. For a constant diffusion coefficient $c = 1$, solving the diffusion equation (2.1) is analogous to convolving the input image f with a Gaussian of a standard deviation $\sigma = \sqrt{2t}$.

Important cues, such as edges and critical points, are gathered from information distributed over all scales in order to analyze the scene as a whole. One of the problems associated with this approach is that important structural features such as edges are smoothed and blurred along the flow, as the processed image evolves in time. As a consequence, the trajectories of zero crossings of the second derivative, which indicate the locations of edges, vary from scale to scale.

2.2.2 Perona–Malik Nonlinear Diffusion

Perona and Malik (P-M) [6] addressed the issue of edge preservation by using the general divergence diffusion form to construct a nonlinear adaptive denoising process, where diffusion can take place with a spatially varying diffusion coefficient in order to reduce the smoothing effect near edges.

The general diffusion equation, controlled by the gradient magnitude, is of the form:

$$u_t = \mathrm{div}(c(|\nabla u|)\nabla u), \tag{2.2}$$

where in the P-M case, c is a positive decreasing function of the gradient magnitude. Two functions for the diffusion coefficient were proposed:

$$c_{PM1}(|\nabla u|) = \frac{1}{1 + (|\nabla u|/k_{PM})^2}$$

and

$$c_{PM2}(|\nabla u|) = \exp((|\nabla u|/k_{PM})^{-2}).$$

It turns out that both have similar basic properties (positive coefficient, non-convex potentials, ability for some local enhancement of large gradients).

Results obtained with the P-M process paved the way for a variety of PDE-based methods that were applied to various problems in low-level vision (see [7] and references cited therein). Some drawbacks and limitations of the original model have been mentioned in the literature (e.g., [8, 9]). Catte et al. [8] have shown the ill-posedness of the diffusion equation, imposed by using the P-M diffusion coefficients, and proposed a regularized version wherein the coefficient is a function of a smoothed gradient:

$$u_t = \mathrm{div}(c(|\nabla u * g_\sigma|)\nabla u). \tag{2.3}$$

Note that although this formulation solved a deep theoretical problem associated with the Perona–Malik process, the characteristics of this process essentially remained. Weickert et al. [10] investigated the stability of the P-M equation by spatial discretization, and proposed [11] a generalized regularization formula in the continuous domain.

Two other useful diffusion coefficients that fall within the Perona–Malik framework, Eq. (2.2), are the Charbonnier diffusivity [12]:

$$c_{Ch}(|\nabla u|) = \frac{1}{\sqrt{1 + (|\nabla u|/k)^2}}$$

and TV-flow [13], which minimizes the total variation energy of Rudin–Osher–Fatemi [6] without the fidelity term:

$$c_{TV}(|\nabla u|) = \frac{1}{|\nabla u|}.$$

The TV-flow coefficient approaches infinity as $|\nabla u| \to 0$ and the process has some interesting unique properties, as we will see later. It is often approximated by a Charbonnier-type formula: $c_{TV\varepsilon}(|\nabla u|) = \frac{1}{\sqrt{\varepsilon^2 + |\nabla u|^2}}$.

2.2.3 Weickert's Anisotropic Diffusion

Weickert suggested a further generalization, preferring smoothing along the local dominant direction (for each pixel), where a tensor diffusion coefficient is used.

2.2.3.1 Tensor Diffusivity

For oriented flow-like structures, such as fingerprints, truly anisotropic processes are called for. Processes emerging from Eq. (2.2) are controlled by a scalar diffusion coefficient $c(x, y, t)$. This permits a spatially varying process that can also change in time throughout the evolution, but is basically isotropic, that is, locally the process acts the same in all directions (in the regularized version, see [10]). Weickert [14, 15] suggested an effective anisotropic scheme using a tensor diffusivity. The diffusion tensor is derived by manipulation of the eigenvalues of the smoothed structure tensor $\mathcal{J}_\sigma = g_\sigma * (\nabla u_\sigma \nabla u_\sigma^T)$.

This technique results in strong smoothing along edges and low smoothing across them. In relatively homogenous regions without coherent edges, the process approaches linear diffusion with low diffusivity. The semilocal nature of the process may extract information from a neighborhood of radius proportional to σ. This enables completion of interrupted lines and enhances flow-like structures.

2.2.3.2 Characteristics of the Flow

The anisotropic diffusion of Weickert still possesses some important characteristics which ensures stability and well behavior of the evolution (see more in Weickert's book [16]):

- Mean value is preserved
- Existence and uniqueness
- Extremum principle
- Well posedness, continuous dependence on the initial image.
- Gray-level shift invariance
- Translation invariance
- Rotation invariance

We will give some more details on these properties in the next part.

2.2.4 Steady-State Solution

The solution at $t \to \infty$ is called the steady-state solution, after the system reaches an equilibrium state. For steady states, we have

$$u_t = 0.$$

Therefore, for the linear diffusion equation $u_t = \Delta u$ we have

$$\Delta u = 0,$$

which is the Laplace equation. The solution depends on the boundary conditions. In the 1D case, $u_{xx} = 0$ which means the solution is of the form $ax + b$.

Example: Assume we have a function $u(t; x)$ defined on the interval [0,1] with initial conditions $u(t = 0; x) = f(x)$, then a steady-state solution can be computed analytically, depending on the boundary conditions:

1. For Dirichlet boundary conditions $u(t; x = 0) = 0$, $u(t; x = 1) = 2$ the steady state is $u(t = \infty; x) = 2x$.
2. For zero Neumann boundary conditions, $u_x(t; x = 0) = u_x(t; x = 1) = 0$, we have $u(t = \infty; x) = \text{const} = \int_0^1 f(x)dx$.

2.2.5   Inverse Scale Space

Due to the systematic bias or loss of contrast introduced by minimizations with the L^2 fidelity term (like ROF) or by gradient flow techniques (e.g. TV-flow), Osher et al. proposed the *Bregman iteration* [17] which—in the continuous limit—leads to the *inverse scale space flow*

$$\partial_t p(t) = f - u(t), \qquad p(t) \in \partial J(u(t)), \ \ p(0) = 0, \tag{2.4}$$

and avoids the systematic error.

2.3   Optical Flow and Registration

2.3.1   Background

Optical flow is a fundamental problem in computer vision. The general goal is to find a displacement field between two similar images. This is related to different vision tasks, three main ones are

1. *Motion Estimation*. This is the origins of the optical flow problem. When processing and analyzing videos, crucial information is which objects moved and where to. This allows tracking of people, cars or moving targets, analyzing gestures, estimating camera motion, etc.
2. *Image Registration*. Transforming a set of images into a single coordinate system. Very useful in medical imaging, automatic target recognition and for processing satellite images. The registration allows easy integration and comparison of the data. Can be single or multimodal (various sensors or image acquisition technologies).

3. *Stereo and 3D Reconstruction*. Computing the geometry of shapes in the image based on several images, either from multiple cameras or from a sequence taken from a moving camera.

Here, we will focus on the optical flow problem for motion estimation.

2.3.1.1 The Optical Flow Equation

We assume to have a video sequence $f(x, y, t)$, where (x, y) are the spatial coordinates and t is the time. Note that here t stands for actual time of the sequence (and is not an artificial evolution parameter). For simplicity we consider two images, sampled at a time interval dt: $f(x, y, t)$ and $f(x, y, t + dt)$. The optical flow assumption is of *gray-level constancy*. That is - the objects can change location, but their color stays the same. Having a displacement vector $(\Delta x, \Delta y)$ for each pixel, we can write

$$f(x, y, t) = f(x + \Delta x, y + \Delta y, t + dt). \tag{2.5}$$

Taking a first-order Taylor approximation we have

$$f(x+\Delta x, y+\Delta y, t+dt) = f(x, y, t) + \frac{\partial f}{\partial x}\Delta x + \frac{\partial f}{\partial y}\Delta y + \frac{\partial f}{\partial t}dt + \text{Higher order terms}. \tag{2.6}$$

Writing the displacement vector as $v = (\frac{\Delta x}{dt}, \frac{\Delta y}{dt}) = (v_1(x, y), v_2(x, y))$ and dividing (2.6) by dt, we get from (2.6) and (2.5) the first-order approximation of the flow, known as the optical flow equation (or optical flow constraint):

$$f_x v_1 + f_y v_2 + f_t = 0. \tag{2.7}$$

Note that this optical flow model is very simple and is fulfilled only partly in real-world scenarios. There are many reasons why in practice this equation will not model well some motions in the image. The main deviations from the intensity constancy assumption are due to

- *Occlusions*. Objects which appear in the first image can be occluded in the second, and vice versa. Therefore pixel intensities can "disappear" or "pop up" between frames.
- *Lighting*. The lighting changes in the scene spatially, therefore the pixel intensity changes when the object moves. Also lighting can change in time (turning on/off lights, clouds moving etc.).
- *Shadows*. This is a specific part of light change, which is very common and is hard to avoid. For instance, the shadow of a person casted on the floor is moving, although the floor, naturally, is still.
- *Normal angle change*. The angle of the normal of a moving object changes with respect to the camera. Thus the intensity can change.

- *Scaling*. The resolution and camera parameters are not taken into account. For example, when a textured object approaches the camera—a single pixel can "become" two pixels in the next frame with different intensities.
- *Movement at boundaries*. For still camera—moving objects come into the frame and out of the frame. For moving camera, or zooming in or out, in addition, the entire scene can change at the boundaries (also of still objects).

Under-determined problem: Another problem of solving the optical flow problem is that the optical flow equation (2.7) gives a single equation for every pixel, whereas the vector flow v has two components for every pixel. Therefore, the problem is under-determined and we need to resort to spatial regularity constraints.

The aperture problem: For constant regions in the image, we get a large ambiguity, where one cannot determine local motion. The equations become degenerate, with many possible solutions. In these cases one needs to resort to additional assumptions of significant spatial regularity.

2.3.2 Early Attempts for Solving the Optical Flow Problem

There have been two different distinct directions to solve the problem, a local approach suggested by Lucas and Kanade in [18] and a global approach suggested by Horn and Schunk in [19].

Lucas–Kanade: They assumed the flow is essentially constant in a local neighborhood of the pixel. For instance, for two pixels, one can write Eq. (2.7) for each pixel and assume (v_1, v_2) are the same. Therefore the problem can be theoretically solved. To increase robustness they increased the window size and solved a least square problem. In the case of large flat regions—there was still large ambiguity.

Horn–Schunck: They suggested a global variational approach. To solve the optical flow problem the following functional was minimized:

$$E_{HS} = \int_{\Omega} (f_x v_1 + f_y v_2 + f_t)^2 + \alpha(|\nabla v_1|^2 + |\nabla v_2|^2)dx. \qquad (2.8)$$

This problem can be solved numerically, for example by a gradient descent based on the Euler–Lagrange equations with respect to v_1 and v_2:

$$\begin{aligned} f_x(f_x v_1 + f_y v_2 + f_t) - \alpha \Delta v_1 = 0, \\ f_y(f_x v_1 + f_y v_2 + f_t) - \alpha \Delta v_2 = 0. \end{aligned} \qquad (2.9)$$

As noted in [20], this type of solution has a filling-in effect: at location with $|\nabla f| \approx 0$ no reliable local flow estimate is possible (aperture problem) but the regularizer $|\nabla v_1|^2 + |\nabla v_2|^2$ fills in information from the neighborhood, resulting in a dense flow. This is a clear advantage over local methods.

In the next section, we will follow this approach.

2.3.3   Modern Optical Flow Techniques

There have been many attempts to solve the various problems encountered in optical flow, using various regularizers and different models, for some of them see [20–25].

We will present in more depth one of the more successful solutions, which is simple enough to understand, yet performs very well and was considered state of the art for a long duration since its publication in 2004.

2.3.3.1   Brox et al Model

This model was suggested by Brox–Bruhn–Papenberg–Weickert (BBPW) in [26]. The total energy is composed of the standard general two terms:

$$E_{BBPW}(v) = E_{Data} + \alpha E_{Smooth}. \tag{2.10}$$

Let us define $\mathbf{w} = (v_1, v_2, 1)$, $\mathbf{x} = (x, y, t)$ and for simplicity, we assume the difference in time between frames is a unit time $dt = 1$. We can also define a spatiotemporal gradient by

$$\nabla_3 = (\partial_x, \partial_y, \partial_t).$$

Let

$$\Psi(s^2) = \sqrt{s^2 + \varepsilon^2}.$$

The model is essentially an L^1 model of the data and smoothness terms.

The data term consists of gray-level constancy as well as gradient constancy assumption:

$$E_{Data}(v) = \int_\Omega \Psi(|f(\mathbf{x} + \mathbf{w}) - f((\mathbf{x}))|^2 + \gamma|\nabla f(\mathbf{x} + \mathbf{w}) - \nabla f((\mathbf{x}))|^2)d\mathbf{x}. \tag{2.11}$$

We assume piecewise smoothness of the flow field. Therefore a total variation approximation of the flow field is minimized:

$$E_{Smooth}(v) = \int_\Omega \Psi(|\nabla_3 v_1|^2 + |\nabla_3 v_2|^2)d\mathbf{x}. \tag{2.12}$$

For the sophisticated numerical methods for solving this model—see the paper [26].

2.4 Segmentation and Clustering

2.4.1 The Goal of Segmentation

The goal of segmentation is generally to generate a higher level representation / understanding of the image by partitioning it into homogeneous parts (or combining similar pixels to a set). The idea behind it is that objects are often combined of just a few segments. A good segmentation of the image often helps higher levels of computer vision algorithms.

The segmentation problem is very hard to solve. It is also not completely well defined, as homogeneous regions can be homogeneous not only in color but also in textures, patterns, orientations, and other features. For specific applications, often some previous segmentation examples or direction from the user are needed to obtain meaningful results.

In this part, we survey some classical variational segmentation models such as Mumford–Shah [27], Chan–Vese [28], and Geodesic-Active-Contours [29].

2.4.2 Mumford–Shah

Mumford and Shah suggested in [27] the following model to solve the segmentation problem:

$$E_{MF}(u, L) = \int_{\Omega-L} (u - f)^2 dx + \alpha \int_{\Omega-L} |\nabla u|^2 dx + \beta \int_L ds, \qquad (2.13)$$

where, as usual, Ω is the image domain and f is the input image. Here, we have a new concept $L \subset \Omega$ which is the set of discontinuities. The minimization is performed jointly over u and L. In other words, one tries to meet the following requirements:

- u is a piecewise smooth function which approximates f well.
- u is smooth (in the H^1 sense) everywhere but on the set of discontinuities L.
- The set L should be as short as possible.

This functional is very hard to minimize numerically. Ambrosio and Tortorelli therefore proposed in [30] an approximated functional with the following formulation:

$$E_{AT}(u, v) = \int_{\Omega} (u-f)^2 dx + \alpha \int_{\Omega} v^2 |\nabla u|^2 dx + \beta \int_{\Omega} \left(\varepsilon |\nabla v|^2 + \frac{1}{4\varepsilon}(v-1)^2 \right) dx. \qquad (2.14)$$

Here a new variable v was introduced, which has a similar role to the discontinuity set L. It has the following properties:

- It is close to 1 whenever u is sufficiently smooth.

- It is close to 0 near large gradients of u.
- It is smooth (in the H^1 sense).

As ε approaches 0, we get a more faithful approximation to the Mumford–Shah functional. More can be found in Theorem 4.2.7 and 4.2.8 in [31] and in details in [30, 32]. A more sophisticated and efficient way to solve the M-S functional was suggested by Pock et al. in [33].

2.4.3 Chan–Vese Model

Chan and Vese in [28] suggested a binary model for object and background. The object and background are assumed to have more or less homogeneous color, or homogeneous average color for patterns and textures. Their concise formulation for the problem is to minimize the following energy:

$$E_{CV}(C, c_1, c_2) = \int_{\text{inside}(C)} (f - c_1)^2 dx + \int_{\text{outside}(C)} (f - c_2)^2 dx + \mu \int_C ds, \quad (2.15)$$

where the energy is minimized with respect to the closed curve C and the two unknown constants c_1, c_2.

A distinct advantage is the ability to go beyond edges and derivatives and to segment also patterns and points with different concentrations (like the Europe lights map). In the original formulation [28] another energy term, the area inside the curve, was also suggested. In addition, different weights between the energy terms for object and background can be assigned.

The Chan–Vese functional can be viewed as a binary or piecewise constant version of the Mumford–Shah functional. See more details in [31], p. 210.

Note that for a fixed curve C, optimizing for c_1, c_2 is done simply by computing the mean value of f inside and outside the curve, respectively. Let A_C be the set inside C, $|A_C| = \int_{A_C} dx$ is the area of A_C and $|\Omega| - |A_C|$ is the area outside A_C. Then

$$\begin{aligned} c_1 &= \tfrac{1}{|A_C|} \int_{A_C} f(x)dx, \\ c_2 &= \tfrac{1}{|\Omega|-|A_C|} \int_{\Omega - A_C} f(x)dx, \end{aligned} \quad (2.16)$$

Algorithm: The general algorithm is

1. Initialize with some curve $C = C_0$.
2. Compute c_1, c_2 according to Eq. (2.16).
3. For fixed c_1, c_2, minimize E_{CV} with respect to the curve C.
4. Repeat stages 2 and 3 until convergence.

A level set formulation was proposed to evolve the curve C and minimize the functional given in (2.15).

2.4.3.1 Convex Model

Chan, Esedoglu, and Nikolova suggested in [34] a convex model that finds a minimizer for the Chan–Vese functional, for a fixed c_1, c_2:

$$E_{CEN}(u, c_1, c_2) = \int_\Omega |\nabla u| dx + \alpha \int_\Omega \left((f(x) - c_1)^2 - (f(x) - c_2)^2 \right) u(x) dx,$$
(2.17)

where $0 \leq u(x) \leq 1$. The segmentation set A_C is recovered by thresholding u:
$A_C := \{x \mid u(x) > 0.5\}$.

Algorithm: The algorithm is similar to the original Chan–Vese model, where one alternates between solving for u and recomputing the constants c_1, c_2. Solving for u, however, is not done using level sets, which are usually used to evolve curves, but using standard variational techniques for evolving functions.

Below we describe the gradient descent method suggested in the original paper [34]: One can introduce a convex penalty function $v(q)$ which is 0 for the interval $q \in [0, 1]$, descends linearly for $q < 0$ and ascends linearly for $q > 1$, such that $v'(q) = -1, q < 0, v'(q) = 1, q > 1$ and 0 otherwise. Then the following gradient descent is used:

$$u_t = \text{div} \left(\frac{\nabla u}{|\nabla u|} \right) - \alpha s(x) - \alpha v'(u),$$
(2.18)

where

$$s(x) = (f(x) - c_1)^2 - (f(x) - c_2)^2.$$

2.4.4 Active Contours

The original active contours model of Kass–Witkin–Terzopoulos [35] inspired many segmentation algorithms. One can visualize their segmentation flow as a process where a band is closing-in around an object.

Let $C(q)$ be a closed curve defined by a parameter $q \in [0, 1]$:

$$C(q) = (C_1(q), C_2(q)), \quad C'(q) = \left(\frac{dC_1}{dq}, \frac{dC_2}{dq} \right).$$

The assumption of closed curve means $C(0) = C(1)$. For s the gradient magnitude, we define an inverse edge detector (attains low values on edges) by

$$g(s) = \frac{1}{1 + s^2}.$$

(this is similar to the P-M diffusion coefficient). Then the following active contour energy can be defined:

$$E_{AC}(C) = \int_0^1 |C'(q)|^2 dq + \beta \int_0^1 |C''(q)|^2 dq + \alpha \int_0^1 g^2(|\nabla f(C(q))|) dq. \quad (2.19)$$

The first two terms are called *internal energy* terms and the last term is an *external energy* term. The idea is that the "band" will be

- Short (first term penalizes length).
- Not too elastic (second term penalizes high curvature).
- On edges of the input image (third term penalizes smooth regions).

Some drawbacks of the method:

1. The functional depends on the curve parametrization (is not intrinsic).
2. The method assumes there is exactly one object. The curve cannot change topology.
3. One needs to solve a fourth-order PDE, which is hard to implement numerically.

Using the foundations of the level set formulation to evolve curves and surfaces in a stable manner, a new level set formulation was proposed by Caselles, Kimmel, and Sapiro called Geodesic-Active-Contours [29]. This became an extremely popular method, with many applications and a comprehensive theory. However this large topic is beyond the scope of this book. For details see the books by Sethian [36], Fedkiw–Osher [37] and Kimmel [38].

The level set formulation [39] is based on the observation that *a curve can be seen as a zero level of a function in a higher dimension*. This is today the standard way of evolving numerically curves and surfaces [40].

Regarding variational segmentation more reading can be found, e.g., in [33, 41–43].

2.5   Patch-Based and Nonlocal Models

2.5.1   Background

One can often better process image pixels by exploiting nonlocal correlations, in addition to the classical local correlations of adjacent pixels. Most notably this has been shown by Efros and Leung for texture synthesis [44] and by Buades–Coll–Morel for denoising [45].

A first suggestion of a nonlocal functional was proposed by Kindermann et al. in [46]. However, no nonlocal operators were suggested and it was very hard to generalized these formulations and to relate them to local functionals.

In [47, 48], Gilboa and Osher proposed a complete nonlocal framework in the continuous setting, derived the basic nonlocal differential operators, and related them to the local setting and to spectral graph theory [49–52].

More advances on this subject can be seen in [53–55]. Here we focus on the paper of [48].

2.5.1.1   Nonlocal Means

In [45], Buades–Coll–Morel suggested the following nonlocal filter for image denoising:

$$NL(u)(x) = \frac{1}{c(x)} \int_\Omega e^{-d_a(u(x),u(y))/h^2} u(y) dy \qquad (2.20)$$

where

$$d_a(u(x), u(y)) = \int_\Omega G_a(t)|u(x+t) - u(y+t)|^2 dt \qquad (2.21)$$

G_a is a Gaussian with standard deviation a, and $c(x)$ is a normalization factor:

$$c(x) = \int_\Omega e^{-d_a(u(x),u(y))/h^2} dy. \qquad (2.22)$$

The corresponding discrete formulation is

$$NL(u)(i) = \sum_j \alpha(i, j) u(j),$$

where

$$\alpha(i, j) = \frac{1}{c(i)} e^{-\|u(B_i)-v(B_j)\|_2^2/h^2}$$

$u(B_i) = (u(k) : k \in B_i)$, B_i is a small ball around pixel i. In practice, one usually uses a square patch of size 5×5 or 7×7 around the center pixel.

2.5.2   Graph Laplacian

We will see below that the operators can be related to similar ones on graphs, and specifically to the well-known graph Laplacian.

Let $G = (V, E)$ be a connected undirected weighted graph with (a finite set of) vertices (nodes) V and edges E. To each edge $e_{kl} \in E$ between nodes k and l, a corresponding weight $w_{kl} \in E$ is defined. The weights are nonnegative and symmetric: $w_{kl} \geq 0$, $w_{kl} = w_{lk}$. We assume that a discrete function u is defined on the nodes of the graph and denote by $u(k) \in V$ the value of u at node k. The (weighted) *graph Laplacian* is

$$\Delta_G(u(k)) := \sum_{l \in \mathcal{N}_k} w_{kl}(u(l) - u(k)), \quad k, l \in V, \qquad (2.23)$$

where $l \in \mathcal{N}_k$ is the set of nodes with edges connected to k. Note that the weights w_{kl} are usually data driven. That is, there is a preprocessing stage of building the graphs (computing the weights), where w_{kl} are derived based on some affinity function. For example, to set the weights between two pixels in the image one can compute an affinity which is inversely proportional to the distance between their surrounding patches. Note that we define here the Laplacian with an *opposite sign* to the usual graph theoretic definition so it will coincide with the continuous definition.

2.5.3 A Nonlocal Mathematical Framework

2.5.3.1 Basic Operators

Let $\Omega \subset \mathbb{R}^n$, $x \in \Omega$, $u(x)$ a real function $u : \Omega \to \mathbb{R}$. We extend the notion of derivatives to a nonlocal framework by the following definition:

$$\partial_y u(x) := \frac{u(y) - u(x)}{\tilde{d}(x, y)}, \quad y, x \in \Omega,$$

where $0 < \tilde{d}(x, y) \le \infty$ is a positive measure defined between points x and y. To keep with standard notations related to graphs, we define the weights as

$$w(x, y) = \tilde{d}^{-2}(x, y).$$

Thus $0 \le w(x, y) < \infty$. The weights are symmetric, that is $w(x, y) = w(y, x)$. The nonlocal derivative can be written as

$$\partial_y u(x) := (u(y) - u(x))\sqrt{w(x, y)}. \tag{2.24}$$

The nonlocal gradient $\nabla_w u(x) : \Omega \to \Omega \times \Omega$ is defined as the vector of all partial derivatives:

$$(\nabla_w u)(x, y) := (u(y) - u(x))\sqrt{w(x, y)}, \quad x, y \in \Omega. \tag{2.25}$$

Vectors are denoted as $\mathbf{v} = v(x, y) \in \Omega \times \Omega$. The standard L^2 inner product is used for functions

$$\langle u_1, u_2 \rangle := \int_\Omega u_1(x) u_2(x) dx.$$

For vectors we define a dot product

$$(\mathbf{v_1} \cdot \mathbf{v_2})(x) := \int_\Omega v_1(x, y) v_2(x, y) dy,$$

and an inner product

$$\langle \mathbf{v}_1, \mathbf{v}_2 \rangle := \langle \mathbf{v}_1 \cdot \mathbf{v}_2, 1 \rangle = \int_{\Omega \times \Omega} v_1(x, y) v_2(x, y) dx dy.$$

The magnitude of a vector is

$$|\mathbf{v}|(x) := \sqrt{\mathbf{v}_1 \cdot \mathbf{v}_1} = \sqrt{\int_{\Omega} v(x, y)^2 dy}.$$

With the above inner products the nonlocal divergence $\mathrm{div}_w \, \mathbf{v}(x) : \Omega \times \Omega \to \Omega$ is defined as the adjoint of the nonlocal gradient:

$$(\mathrm{div}_w \, \mathbf{v})(x) := \int_{\Omega} (v(x, y) - v(y, x)) \sqrt{w(x, y)} dy. \tag{2.26}$$

The Laplacian can now be defined by

$$\Delta_w u(x) := \frac{1}{2} \mathrm{div}_w (\nabla_w u(x)) = \int_{\Omega} (u(y) - u(x)) w(x, y) dy. \tag{2.27}$$

Note that in order to get the standard Laplacian definition which relates to the graph Laplacian, we need a factor of $1/2$.

2.5.3.2   Some Properties

Most of the properties involving a double integral can be shown by expanding an integral of the form $\int_{\Omega \times \Omega} f(x, y) dx dy$ to $\frac{1}{2} \int_{\Omega \times \Omega} (f(x, y) + f(y, x)) dx dy$, changing the order of integration and using the symmetry property of w. We give an example showing the adjoint relation

$$\langle \nabla_w u, \mathbf{v} \rangle = \langle u, - \mathrm{div}_w \, \mathbf{v} \rangle, \tag{2.28}$$

$$\begin{aligned}
\langle \nabla_w u, \mathbf{v} \rangle &= \int_{\Omega \times \Omega} (u(y) - u(x)) \sqrt{w(x, y)} v(x, y) dx dy \\
&= \frac{1}{2} \int_{\Omega \times \Omega} \left[(u(y) - u(x)) \sqrt{w(x, y)} v(x, y) + (u(x) - u(y)) \sqrt{w(y, x)} v(y, x) \right] dx dy \\
&= \frac{1}{2} \int_{\Omega \times \Omega} [u(y)(v(x, y) - v(y, x)) - u(x)(v(x, y) - v(y, x))] \sqrt{w(x, y)} dx dy \\
&= \frac{1}{2} \int_{\Omega \times \Omega} [u(x)(v(y, x) - v(x, y)) - u(x)(v(x, y) - v(y, x))] \sqrt{w(x, y)} dx dy \\
&= \int_{\Omega} u(x) \left(- \int_{\Omega} (v(x, y) - v(y, x)) \sqrt{w(x, y)} dy \right) dx.
\end{aligned}$$

"Divergence theorem":

$$\int_{\Omega} \mathrm{div}_w \, \mathbf{v} dx = 0. \tag{2.29}$$

The Laplacian is self-adjoint

$$\langle \Delta_w u, u \rangle = \langle u, \Delta_w u \rangle \tag{2.30}$$

and negative semi-definite

$$\langle \Delta_w u, u \rangle = -\langle \nabla_w u, \nabla_w u \rangle \leq 0. \tag{2.31}$$

We can also formulate a nonlocal (mean) curvature:

$$\begin{aligned}
\kappa_w &:= \text{div}_w \left(\frac{\nabla_w u}{|\nabla_w u|} \right) \\
&= \int_\Omega (u(y) - u(x)) w(x, y) \left(\frac{1}{|\nabla_w u|(x)} + \frac{1}{|\nabla_w u|(y)} \right) dy,
\end{aligned} \tag{2.32}$$

where

$$|\nabla_w u|(q) := \sqrt{\int_\Omega (u(z) - u(q))^2 w(q, z) dz}.$$

2.5.3.3 The Regularizing Functionals

Two types of regularizing nonlocal functionals are proposed. The first type is based on the nonlocal gradient. It is set within the mathematical framework described above. The second type is based on differences, it appears to be easier to implement, where the minimization can be accomplished using graph cut techniques,

The gradient-based functional is

$$\begin{aligned}
J(u) &= \int_\Omega \phi(|\nabla_w u|^2) dx, \\
&= \int_\Omega \phi(\int_\Omega (u(y) - u(x))^2 w(x, y) dy) dx,
\end{aligned} \tag{2.33}$$

where $\phi(s)$ is a positive function, convex in $\sqrt{s}$ with $\phi(0) = 0$.

The difference-based functional is

$$J_a(u) = \int_{\Omega \times \Omega} \phi((u(y) - u(x))^2 w(x, y)) dy dx. \tag{2.34}$$

The variation with respect to u (Euler–Lagrange) of (2.33) is

$$\partial_u J(u) = -2 \int_\Omega (u(y) - u(x)) w(x, y) (\phi'(|\nabla_w u|^2(x)) + \phi'(|\nabla_w u|^2(y))) dy, \tag{2.35}$$

where $\phi'(s)$ is the derivative of ϕ with respect to s. This can be written more concisely as

$$\partial_u J(u) = -2 \, \text{div}_w \left(\nabla_w u \phi'(|\nabla_w u|^2(x)) \right).$$

The variation with respect to u of (2.34) is

$$\partial_u J_a(u) = -4 \int_\Omega (u(y) - u(x))w(x,y)\phi'((u(y) - u(x))^2 w(x,y))dy. \quad (2.36)$$

Note that for the quadratic case $\phi(s) = s$ the functionals (2.33) and (2.34) coincide (and naturally so do Eqs. (2.35) and (2.36)).

As an example, for total variation, $\phi(s) = \sqrt{s}$, Eq. (2.33) becomes

$$J_{NL-TV}(u) = \int_\Omega |\nabla_w u|dx = \int_\Omega \sqrt{\int_\Omega (u(y) - u(x))^2 w(x,y)dy}dx \quad (2.37)$$

whereas Eq. (2.34) becomes

$$J_{NL-TVa}(u) = \int_{\Omega \times \Omega} |u(x) - u(y)|\sqrt{w(x,y)}dydx \quad (2.38)$$

The above functionals correspond in the local two-dimensional case to the isotropic TV

$$J_{TV}(u) = \int_\Omega |\nabla u|dx = \int_\Omega \sqrt{u_{x_1}^2 + u_{x_2}^2}dx$$

and to the anisotropic TV

$$J_{TVa}(u) = \int_\Omega (|u_{x_1}| + |u_{x_2}|)dx.$$

2.5.4 Basic Models

Classical PDE's can have their nonlocal generalization:

Nonlocal diffusion:

$$u_t(x) = \Delta_w u(x), \quad u_{t=0} = f(x), \quad (2.39)$$

where $\Delta_w u(x)$ is the nonlocal Laplacian defined in (2.27).

Nonlocal TV-flow:

$$u_t(x) = -\partial J_{NL-TV}(u(x)), \quad u_{t=0} = f(x), \quad (2.40)$$

where $J_{NL-TV}(u)$ is defined in (2.37).

One can also define regularization models, based for instance on the NL-TV.

Nonlocal ROF:

$$J_{NL-TV}(u) + \alpha\|f - u\|_{L^2}^2. \tag{2.41}$$

Nonlocal TV-L1: Another very important model following [3] is the extension of $TV - L^1$ to a nonlocal version:

$$J_{NL-TV}(u) + \alpha\|f - u\|_{L^1}. \tag{2.42}$$

An interesting application of texture regularization can be shown using $TV - L^1$. It can both detect and remove anomalies or irregularities from images, and specifically textures.

Inpainting: Follows the local TV-inpainting model of [56]:

$$J_{NL-TV}(u) + \int_\Omega \alpha(x)(f - u)^2 dx, \tag{2.43}$$

with $\alpha(x) = 0$ in the inpainting region and $\alpha(x) = c$ in the rest of the image.

More details on the discretization and examples of numerical methods to solve such functionals appear in Appendix A.5.

References

1. A. Tikhonov, Solution of incorrectly formulated problems and the regularization method. Sov. Math. Dokl. **4**, 1035–1038 (1963)
2. L. Rudin, S. Osher, E. Fatemi, Nonlinear total variation based noise removal algorithms. Physica D **60**, 259–268 (1992)
3. M. Nikolova, A variational approach to remove outliers and impulse noise. JMIV **20**(1–2), 99–120 (2004)
4. T.F. Chan, S. Esedoglu, Aspects of total variation regularized l 1 function approximation. SIAM J. Appl. Math. **65**(5), 1817–1837 (2005)
5. B.M. Ter Haar Romeny, Introduction to scale-space theory: multiscale geometric image analysis. in *First International Conference on Scale-Space theory* (Citeseer, 1996)
6. P. Perona, J. Malik, Scale-space and edge detection using anisotropic diffusion. PAMI **12**(7), 629–639 (1990)
7. B.M. Ter Haar Romeny (ed.), *Geometry Driven Diffusion in Computer Vision* (Kluwer Academic Publishers, Dordrecht, 1994)
8. F. Catté, P.-L. Lions, J.-M. Morel, T. Coll, Image selective smoothing and edge detection by nonlinear diffusion. SIAM J. Numer. Anal. **29**(1), 182–193 (1992)
9. R.T. Whitaker, S.M. Pizer, A multi-scale approach to nonuniform diffusion. CVGIP Image Underst. **57**(1), 99–110 (1993)
10. J. Weickert, B. Benhamouda, A semidiscrete nonlinear scale-space theory and its relation to the perona malik paradox, in *Advances in Computer Vision* (Springer, Berlin, 1997), pp. 1–10
11. E. Radmoser, O. Scherzer, J. Weickert, Scale-space properties of nonstationary iterative regularization methods. J. Vis. Commun. Image Represent. **8**, 96–114 (2000)
12. P. Charbonnier, L. Blanc-Feraud, G. Aubert, M. Barlaud, Two deterministic half-quadratic regularization algorithms for computed imaging. in *Proceedings of the IEEE International Conference ICIP '94*, vol. 2 (1994), pp. 168–172

13. F. Andreu, C. Ballester, V. Caselles, J.M. Mazón, Minimizing total variation flow. Differ. Integral Equ. **14**(3), 321–360 (2001)
14. J. Weickert, Coherence-enhancing diffusion filtering. Int. J. Comput. Vis. **31**(2–3), 111–127 (1999)
15. J. Weickert, Coherence-enhancing diffusion of colour images. IVC **17**, 201–212 (1999)
16. J. Weickert, *Anisotropic Diffusion in Image Processing* (Teubner-Verlag, Germany, 1998)
17. S. Osher, M. Burger, D. Goldfarb, J. Xu, W. Yin, An iterative regularization method for total variation based image restoration. SIAM J. Multiscale Model. Simul. **4**, 460–489 (2005)
18. B.D. Lucas, T. Kanade et al., An iterative image registration technique with an application to stereo vision. IJCAI **81**, 674–679 (1981)
19. B.K.P. Horn, B.G. Schunck. Determining optical flow. Artif. Intell. **17**(1), 185–203 (1981)
20. A. Bruhn, J. Weickert, C. Schnörr, Lucas/kanade meets horn/schunck: combining local and global optic flow methods. Int. J. Comput. Vis. **61**(3), 211–231 (2005)
21. G. Aubert, R. Deriche, P. Kornprobst, Computing optical flow via variational techniques. SIAM J. Appl. Math. **60**(1), 156–182 (1999)
22. C. Zach, T. Pock, H. Bischof, A duality based approach for realtime TV-L 1 optical flow. Pattern Recognit. 214–223 (2007)
23. S. Baker, D. Scharstein, J.P. Lewis, S. Roth, M.J. Black, R. Szeliski. A database and evaluation methodology for optical flow. Int. J. Comput. Vis. **92**(1), 1–31 (2011)
24. D. Sun, S. Roth, M.J. Black, Secrets of optical flow estimation and their principles. in *IEEE Conference on Computer Vision and Pattern Recognition (CVPR), 2010* (IEEE, 2010), pp. 2432–2439
25. D. Cremers, S. Soatto, Motion competition: a variational approach to piecewise parametric motion segmentation. Int. J. Comput. Vis. **62**(3), 249–265 (2005)
26. T. Brox, A. Bruhn, N. Papenberg, J. Weickert, High accuracy optical flow estimation based on a theory for warping, in *Computer Vision-ECCV 2004* (Springer, Berlin, 2004), pp. 25–36
27. D. Mumford, J. Shah, Optimal approximations by piece-wise smooth functions and assosiated variational problems. Commun. Pure Appl. Math. **42**, 577–685 (1989)
28. T. Chan, L. Vese, Active contours without edges. IEEE Trans. Image Process. **10**(2), 266–277 (2001)
29. V. Caselles, R. Kimmel, G. Sapiro, Geodesic active contours. Int. J. Comput. Vis. **22**(1), 61–79 (1997)
30. L. Ambrosio, V.M. Tortorelli, Approximation of functional depending on jumps by elliptic functional via t-convergence. Commun. Pure Appl. Math. **43**(8), 999–1036 (1990)
31. G. Aubert, P. Kornprobst, in *Mathematical Problems in Image Processing*, Applied Mathematical Sciences, vol 147 (Springer, Berlin, 2002)
32. A. Braides, *Approximation of Free-Discontinuity Problems* (Springer, Berlin, 1998)
33. T. Pock, D. Cremers, H. Bischof, A. Chambolle, An algorithm for minimizing the mumford-shah functional. in *IEEE 12th International Conference on Computer Vision, 2009* (IEEE, 2009), pp. 1133–1140
34. T.F. Chan, S. Esedoglu, M. Nikolova, Algorithms for finding global minimizers of image segmentation and denoising models. SIAM J. Appl. Math. **66**(5), 1632–1648 (2006)
35. M. Kass, A. Witkin, D. Terzopoulos, Snakes: active contour models. Int. J. Comput. Vis. **1**(4), 321–331 (1987)
36. J.A. Sethian, *Level Set Methods and Fast Marching Methods: Evolving Interfaces in Computational Geometry, Fluid Mechanics, Computer vision, and Materials Science* (Cambridge university press, Cambridge, 1999)
37. S. Osher, R. Fedkiw, *Level Set Methods and Dynamic Implicit Surfaces* (Springer, Berlin, 2002)
38. R. Kimmel, *Numerical Geometry of Images: Theory, Algorithms, and Applications* (Springer, Berlin, 2012)
39. S. Osher, J.A. Sethian, Fronts propagating with curvature dependent speed-algorithms based on Hamilton-Jacobi formulations. J. Comput. Phys. **79**, 12–49 (1988)
40. S. Osher, N. Paragios (eds.), *Geometric Level Set Methods in Imaging, Vision, and Graphics* (Springer, Berlin, 2003)

41. D. Cremers, M. Rousson, R. Deriche, A review of statistical approaches to level set segmentation: integrating color, texture, motion and shape. Int. J. Comput. Vis. **72**(2), 195–215 (2007)
42. J. Lie, M. Lysaker, X.-C. Tai, A binary level set model and some applications to mumford-shah image segmentation. IEEE Trans. Image Process. **15**(5), 1171–1181 (2006)
43. N. Paragios, R. Deriche, Geodesic active regions and level set methods for supervised texture segmentation. Int. J. Comput. Vis. **46**(3), 223–247 (2002)
44. A.A. Efros, T.K. Leung, Texture synthesis by non-parametric sampling. ICCV **2**, 1033–1038 (1999)
45. A. Buades, B. Coll, J.-M. Morel, A review of image denoising algorithms, with a new one. SIAM Multiscale Model. Simul. **4**(2), 490–530 (2005)
46. S. Kindermann, S. Osher, P. Jones, Deblurring and denoising of images by nonlocal functionals. SIAM Multiscale Model. Simul. **4**(4), 1091–1115 (2005)
47. G. Gilboa, S. Osher, Nonlocal linear image regularization and supervised segmentation. SIAM Multiscale Model. Simul. **6**(2), 595–630 (2007)
48. G. Gilboa, S. Osher, Nonlocal operators with applications to image processing. SIAM Multiscale Model. Simul. **7**(3), 1005–1028 (2008)
49. F. Chung, *Spectral Graph Theory*, vol. 92, CBMS Regional Conference Series in Mathematics (American Mathematical Society, Providence, 1997)
50. S. Bougleux, A. Elmoataz, M. Melkemi, Discrete regularization on weighted graphs for image and mesh filtering. in *1st International Conference on Scale Space and Variational Methods in Computer Vision (SSVM)*. Lecture Notes in Computer Science, vol. 4485 (2007), pp. 128–139
51. D. Zhou, B. Scholkopf, Regularization on discrete spaces. in *Pattern Recognition, Proceedings of the 27th DAGM Symposium* (Berlin, Germany, 2005), pp. 361–368,
52. A.D. Szlam, M. Maggioni, Jr. J.C. Bremer, R.R. Coifman, Diffusion-driven multiscale analysis on manifolds and graphs: top-down and bottom-up constructions. in *SPIE* (2005)
53. C. Kervrann, J. Boulanger, Optimal spatial adaptation for patch-based image denoising. IEEE Trans. Image Process. **15**(10), 2866–2878 (2006)
54. G. Peyré, S. Bougleux, L. Cohen, Non-local regularization of inverse problems, *Computer Vision–ECCV 2008* (Springer, Berlin, 2008), pp. 57–68
55. Y. Lou, X. Zhang, S. Osher, A. Bertozzi, Image recovery via nonlocal operators. J. Sci. Comput. **42**(2), 185–197 (2010)
56. T.F. Chan, J. Shen, Mathematical models of local non-texture inpaintings. SIAM J. Appl. Math. **62**(3), 1019–1043 (2001)

Chapter 3
Total Variation and Its Properties

3.1 Strong and Weak Definitions

The total variation functional is essentially understood as the L^1 norm of the gradient, which for smooth functions can be written in (the more straightforward) strong sense as

$$J_{TV}(u) = \int_{\Omega} |\nabla u(x)| \, dx. \tag{3.1}$$

This formulation is very easy to understand, but it assumes implicitly that u is differentiable. A main advantage of TV, however, is to handle well also discontinuous functions. Thus TV is finite also for non-smooth functions which are in the space of bounded variations BV. A canonical example is a step function

$$u_{step}(x) = \begin{cases} 0 & \text{for } x < 0, \\ 1 & \text{for } x \geq 0 \end{cases}$$

for which $J_{TV}(u_{step}) = 1$. Thus discontinuous functions, such as a step, can have in some cases very low TV value. Note that if one uses quadratic regularizers, such as the Dirichlet energy $J_D(u) = \int_{\Omega} |\nabla u(x)|^2 \, dx$, we get $J_D(u_{step}) = \infty$.

A step is not differentiable, therefore how can TV be computed? There are several ways to understand why the TV value of a step is 1. In 1D, where Ω is the real line, (3.1) translates to

$$J_{TV-1D}(u) = \int_{-\infty}^{\infty} |u_x| \, dx.$$

From an engineering standpoint, one can take the derivative of a step as a Dirac delta function $\delta(x)$, which is nonnegative everywhere to have $\int_{-\infty}^{\infty} \delta(x) dx = 1$. Another way is to approximate a step by $\hat{u}(x)$, a sharp smooth (differential) slope of width ε and since it is nonnegative we can remove the absolute value to get

© Springer International Publishing AG, part of Springer Nature 2018
G. Gilboa, *Nonlinear Eigenproblems in Image Processing and Computer Vision*, Advances in Computer Vision and Pattern Recognition, https://doi.org/10.1007/978-3-319-75847-3_3

$$\int_{-\infty}^{\infty} \hat{u}_x dx = \hat{u}|_{-\infty}^{\infty} = 1 - 0 = 1.$$

The formal way, is to define TV in the weak sense, that is by integrating u multiplied by a test function z (which is differentiable), in the following way

$$J_{TV}(u) = \sup_{\substack{z \in C_c^{\infty}, \\ \|z\|_{L^{\infty}(\Omega)} \le 1}} \int_{\Omega} u(x) \text{ div } z(x) \, dx. \tag{3.2}$$

In this case, the test function z is a vector field which is infinitely differentiable and of compact support. Its magnitude $|z_i|$ is bounded by 1. We will later see that a deep understanding of the structure of z enables us to find TV eigenfunctions and to characterize their geometrical properties.

3.2   Co-area Formula

We now show a fundamental relationship between length and TV. The perimeter of a characteristic function of a shape A, χ_A is denoted by $\text{Per}(\chi_A)$ (the length of the boundary ∂A).

Let us define a characteristic function $E_h(u) = 1$ if $u > h$ and 0 otherwise. The co-area formula is

$$|u|_{TV} = \int_{-\infty}^{\infty} \text{Per}(E_h(u)) dh.$$

A proof can be found for instance in Sect. 2.2.3 of [1].

3.3   Definition of BV

Let Ω be a bounded open subset of $\mathbb{R}^N$ and let $u \in L^1(\Omega)$. Let $\varphi = (\varphi_1, .. \varphi_N)$ be a continuously differentiable function with compact support in Ω (belongs to the space $C_0^1(\Omega)^N$). We denote

$$|\varphi|_{L^{\infty}} = \sup_x \sqrt{\sum_{i=1}^{N} \varphi_i^2(x)}.$$

We define the total variation in the distributional sense as

$$\int_{\Omega} |Du| = \sup_{\varphi} \left\{ \int_{\Omega} u \text{ div } \varphi dx, \ |\varphi|_{L^{\infty}} \le 1 \right\} \tag{3.3}$$

We refer to Du as the distributional gradient.

The space $BV(\Omega)$ is defined as the space of functions of bounded variation:

$$BV(\Omega) = \left\{ u \in L^1(\Omega); \int_\Omega |Du| < \infty \right\}. \tag{3.4}$$

$BV(\Omega)$ is a Banach space endowed with the norm $\|u\|_{BV(\Omega)} = \|u\|_{L^1} + \int_\Omega |Du|$.

3.4 Basic Concepts Related to TV

3.4.1 Isotropic and Anisotropic TV

TV is essentially the L^1 of the gradient magnitude. The gradient magnitude can be defined in several ways. The two most common ones are using the Euclidean norm and the L^1 norm. In the Euclidean 2D case we have

$$|\nabla u|_2 = \sqrt{(u_x)^2 + (u_y)^2}.$$

In the L^1 case we have

$$|\nabla u|_1 = |u_x| + |u_y|.$$

The Euclidean $2 - norm$ case is invariant to rotation of the coordinate system and hence is referred to as *isotropic*, and TV defined by it is *isotropic TV*. For the later $1 - norm$ case, we can clearly see that the length depends on the coordinate angle. For instance, a vector $(u_x, u_y) = (1, 0)$ will have $|\nabla u|_1 = 1$, however if we rotate the system (or image) by 45 degrees we will get $|\nabla u|_1 = \sqrt{2}$. In the latter formulation, TV is separable in any dimension with respect to the coordinate system and is thus easier to compute

$$TV_{anisotropic}(u) = \int_\Omega |\nabla u|_1 dxdy = \int_\Omega (|u_x| + |u_y|)dxdy.$$

However, it produces some artifacts, as it tends to align the objects to the axes. As we will see later, in terms of eigenfunctions, the isotropic TV has disk-like shapes as eigenfunctions, whereas the anisotropic TV has rectangular shapes.

3.4.2 ROF, TV-L1, and TV Flow

There are several ways to use the TV functional. For denoising the ROF model [2] of Rudin–Osher–Fatemi is a fundamental denoising method, which works best for signals which are approximately piecewise constant, with additive white Gaussian noise,

$$E_{ROF}(u,f) = J_{TV}(u) + \lambda\|f - u\|_{L^2}^2. \tag{3.5}$$

When an estimation of the noise standard deviation σ is at hand, a constrained type optimization can be used, where λ is treated as a Lagrange multiplier and is computed by minimizing $J_{TV}(u)$ subject to $\|f - u\|_{L^2}^2 = \sigma^2$.

One can also use a deconvolution formulation to solve inverse problems of a known blur kernel K. The input image f is assumed to be blurred and corrupted by noise $f = K * g + n$, where g is the original image, n is noise and $*$ denotes convolution. The minimization in this case is for the energy

$$E_{ROF-deconv}(u,f,K) = J_{TV}(u) + \lambda\|f - K * u\|_{L^2}^2. \tag{3.6}$$

For the case of removing outliers and impulsive noise (like "salt and pepper") an L^1 fidelity term is used, which is robust to outliers.

$$E_{TV-L1}(u,f) = J_{TV}(u) + \lambda\|f - u\|_{L^1}. \tag{3.7}$$

This energy has some very nice geometric properties and removes structure only according to their size (or diameter) but is invariant to contrast. Some studies and analysis are in [3–5].

Another way of using the total variation functional is to perform a gradient descent evolution, creating a nonlinear scale space flow, often referred to as TV flow [6]

$$u_t = -p, \quad p \in \partial J_{TV}(u), \tag{3.8}$$

where $\partial J_{TV}(u)$ is the subdifferential of TV. The properties of the flow were investigated thoroughly in [6–9]. We will use this flow to construct the spectral TV representation.

For small time step dt, a flow at $u(t + dt)$ can be approximated as applying the proximal operator on the solution at time t, $u(t)$,

$$u(t + dt) = argmin_u\{J_{TV}(u) + \frac{1}{2dt}\|u(t) - u\|_{L^2}^2\}.$$

Note that this is the ROF problem (3.5) with $f = u(t)$ and $\lambda = 1/(2dt)$. This can be understood as an implicit time step discretization, by viewing the Euler–Lagrange equation

$$0 \in \partial J_{TV}(u) + \frac{1}{dt}(u - u(t)).$$

In some settings, such as discrete 1D, it was shown [10, 11] that the flow and minimization are equivalent for any dt.

Finite extinction time. An important property of the flow and for the minimizations is the finite extinction time. For the flow, we have a time T for which for any $t \geq T$ we reach a steady state where no change takes place, $u_t = 0$. In this case, u is in the nullspace of J_{TV} and is simply a constant (where the Neumann boundary conditions yield that the constant is simply the mean value of the initial condition f).

For the ROF problem, a similar phenomenon occurs, there is a threshold λ_T for which for any $\lambda \leq \lambda_T$, the solution of the minimization is a constant.

References

1. T.E. Chan, J. Shen, *Image Processing and Analysis* (SIAM, 2005)
2. L. Rudin, S. Osher, E. Fatemi, Nonlinear total variation based noise removal algorithms. Physica D **60**, 259–268 (1992)
3. M. Nikolova, A variational approach to remove outliers and impulse noise. JMIV **20**(1–2), 99–120 (2004)
4. T.F. Chan, S. Esedoglu, Aspects of total variation regularized l1 function approximation. SIAM J. Appl. Math. **65**(5), 1817–1837 (2005)
5. V. Duval, J.-F. Aujol, Yann Gousseau, The tvl1 model: a geometric point of view. Multiscale Model. Simul. **8**(1), 154–189 (2009)
6. F. Andreu, C. Ballester, V. Caselles, J.M. Mazón, Minimizing total variation flow. Differ. Integral Equ. **14**(3), 321–360 (2001)
7. F. Andreu, V. Caselles, J.I. Dıaz, J.M. Mazón, Some qualitative properties for the total variation flow. J. Funct. Anal. **188**(2), 516–547 (2002)
8. G. Bellettini, V. Caselles, M. Novaga, The total variation flow in R^N. J. Differ. Equ. **184**(2), 475–525 (2002)
9. V. Caselles, K. Jalalzai, M. Novaga, On the jump set of solutions of the total variation flow. Rendiconti del Seminario Matematico della Università di Padova **130**, 155–168 (2013)
10. G. Steidl, J. Weickert, T. Brox, P. Mrzek, M. Welk, On the equivalence of soft wavelet shrinkage, total variation diffusion, total variation regularization, and SIDEs. SIAM J. Numer. Anal. **42**(2), 686–713 (2004)
11. M. Burger, G. Gilboa, M. Moeller, Lina Eckardt, Daniel Cremers, Spectral decompositions using one-homogeneous functionals. SIAM J. Imaging Sci. **9**(3), 1374–1408 (2016)

Chapter 4
Eigenfunctions of One-Homogeneous Functionals

4.1 Introduction

Let us summarize earlier studies concerning the use of nonlinear eigenfunctions in the context of variational methods.

There are several ways to generalize the linear eigenvalue problem $Lu = \lambda u$, where L is a linear operator, to the nonlinear case (for some alternative ways see [1]). A very general and broad definition is

$$T(u) = \lambda Q(u), \tag{4.1}$$

where $T(u)$ is a bounded linear or nonlinear operator defined on an appropriate Banach space $\mathscr{U}$, $\lambda \in \mathbb{R}$ is the eigenvalue and $Q(u)$ can be a linear or nonlinear operator or function of u. In this book, we are mostly concerned with Q as an identity operator, which reduces (4.1) to the following formulation:

$$T(u) = \lambda u. \tag{4.2}$$

We restrict ourselves to the real-valued setting. The operator $T(u)$ is derived based on the derivative of a convex functional.

One main notion in this book is the concept of nonlinear eigenfunctions induced by convex functionals. We refer to u with $\|u\|_2 = 1$ as an *eigenfunction* of J if it admits the following eigenvalue problem:

$$\lambda u \in \partial J(u), \tag{4.3}$$

where $\lambda \in \mathbb{R}$ is the corresponding eigenvalue.

The analysis of eigenfunctions related to non-quadratic convex functionals was mainly concerned with the total variation (TV) regularization. In the analysis of variational TV denoising, i.e., the ROF model from [2], Meyer [3] has shown an explicit

© Springer International Publishing AG, part of Springer Nature 2018
G. Gilboa, *Nonlinear Eigenproblems in Image Processing and Computer Vision*, Advances in Computer Vision and Pattern Recognition, https://doi.org/10.1007/978-3-319-75847-3_4

solution for the case of a disk (an eigenfunction of TV), quantifying explicitly the loss of contrast and advocating the use of $TV - G$ regularization. Within the extensive studies of the TV flow [4–10], eigenfunctions of TV (referred to as *calibrable sets*) were analyzed and explicit solutions were given for several cases of eigenfunction spatial settings. In [11], an explicit solution of a disk for the inverse scale space flow is presented, showing its instantaneous appearance at a precise time point related to its radius and height.

In [12–15], eigenfunctions related to the total generalized variation (TGV) [16] and the infimal convolution total variation (ICTV) functional [17] are analyzed in the case of a bounded domain and their different reconstruction properties on particular eigenfunctions of the TGV are demonstrated theoretically as well as numerically.

Examples of certain eigenfunctions for different extensions of the TV to color images are given in [18].

In [7], Steidl et al. have shown the close relations and equivalence in a 1D discrete setting, of the Haar wavelets to both TV regularization and TV flow. This was later developed for a 2D setting in [19]. The connection between Haar wavelets and TV methods in 1D was made more precise in [20], who indeed showed that the Haar wavelet basis is an orthogonal basis of eigenfunctions of the total variation (with appropriate definition at the domain boundary)—in this case, the Rayleigh principle holds for the whole basis. In the field of morphological signal processing, nonlinear transforms were introduced in [21, 22].

4.2   One-Homogeneous Functionals

We are examining convex functionals of the form $J : \mathbb{R}^n \to \mathbb{R}$ which are absolutely one-homogeneous, that is

$$J(su) = |s| J(u) \qquad \forall s \in \mathbb{R}, \forall u \in \mathscr{X}. \tag{4.4}$$

We can interpret J as a seminorm, respectively, a norm, on an appropriate subspace. The most useful regularizers in image processing, which preserve edges well, are one-homogeneous. This includes all flavors of total variation on grids and graphs as well as higher order functionals, such as total generalized variation (TGV) [16]. Please see Lemmas 1.1–1.6 in Chap. 1 for detailed properties of one-homogeneous functionals and their subgradients.

4.3 Properties of Eigenfunction

It was shown in [23] that when the flow is initialized with an eigenfunction (that is, $\lambda f \in \partial J(f)$) the following solution is obtained:

$$u(t; x) = (1 - \lambda t)^+ f(x), \tag{4.5}$$

where $(q)^+ = q$ for $q > 0$ and 0 otherwise. This means that the shape $f(x)$ is spatially preserved and changes only by contrast reduction throughout time.

A similar behavior (see [20, 23]) can be shown for a minimization problem with the 2 norm, defined as follows:

$$\min_u J(u) + \frac{\alpha}{2} \|f - u\|_2^2. \tag{4.6}$$

In this case, when f is an eigenfunction and $\alpha \in \mathbb{R}^+$ ($\mathbb{R}^+ = \{x \in \mathbb{R} \mid x \geq 0\}$) is fixed, the problem has the following solution:

$$u(x) = \left(1 - \frac{\lambda}{\alpha}\right)^+ f(x). \tag{4.7}$$

In this case also, $u(x)$ preserves the spatial shape of $f(x)$ (as long as $\alpha > \lambda$).

Having a better understanding of properties of the eigenfunctions can assist in the choice of a proper functional for a given image processing task. The behavior of eigenfunctions under some kind of processing is illustrated in a toy example in Fig. 4.1. The TV filtering is based on spectral TV which is explained in Chap. 5.

In Fig. 4.1, a numerical eigenfunction for the discrete TV functional is given (as computed by the Nossek–Gilboa flow described later in Sect. 7.3). It can be seen in Fig. 4.1b that the spectral response $S(t)$ of the eigenfunction approaches a numerical delta. As this is based on a smoothing TV flow, the noise response appears mostly in smaller scales and is well separated from the clean eigenfunction in the transform domain, Fig. 4.1d. Thus, in order to denoise, one performs the nonlinear analog of an ideal low-pass filter with $H(t) = 1$ for $t \geq t_c$ and 0 otherwise (t_c is the cutoff scale, note here that high "frequencies" appear at low t). Denoising an eigenfunction is very suitable for such spectral filtering and maintains full contrast of the clean signal. As can be seen in Fig. 4.1e–g, results compete well with state-of-the-art denoising algorithms such as BM3D [24] or EPLL [25].

Therefore, by having a better understanding of the regularizer and its eigenfunctions, one can enhance the regularization quality by adapting the functionals to fit the class of signals to be processed.

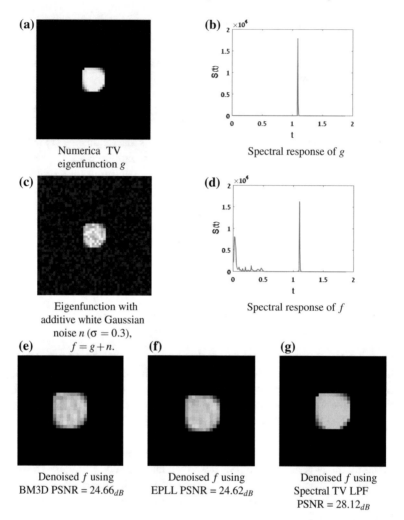

Fig. 4.1 A motivating example for the significance of eigenfunctions in regularization and spectral filtering. It is shown how a regularizer J is very well suited to process an eigenfunction g admitting $\lambda g \in \partial J(g)$. In this case, J is the (discrete) isotropic TV functional. From top left, **a** Eigenfunction g, **b** Its spectral response $S(t)$, **c** Eigenfunction with noise and its spectral response (**d**), performing denoising using: BM3D (**e**), EPLL (**f**), and TV spectral filtering (**g**)

4.4 Eigenfunctions of TV

A highly notable contribution in [6] is the precise geometric characterization of convex sets which are TV eigenfunctions. Let us briefly recall it. Let χ_C, be a characteristic function of the set $C \subset \mathbb{R}^2$, then it admits (4.3), with J the TV functional J_{TV}, if

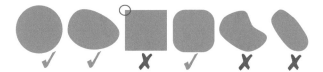

Fig. 4.2 Example of valid and nonvalid shapes as nonlinear eigenfunctions with respect to the TV functional. Smooth enough convex sets, which admit condition (4.8) (marked with a green check), are solutions of the eigenvalue problem (4.3). Shapes with either too high curvature, not convex or too elongated (marked with red X) are not valid

$$\operatorname*{ess\ sup}_{q \in \partial C} \kappa(q) \le \frac{P(C)}{|C|}, \tag{4.8}$$

where C is convex, $\partial C \in C^{1,1}$, $P(C)$ is the perimeter of C, $|C|$ is the area of C, and κ is the curvature. In this case, the eigenvalue is

$$\lambda = \frac{P(C)}{|C|}.$$

See Fig. 4.2 for some examples. Theorem 6 in [6] furthermore classified all possible eigenfunctions arising from the indicator function of a set. Such eigenfunctions must arise from a union of sets C with the above properties, all having the same eigenvalue $\frac{P(C)}{|C|}$ and additionally being sufficiently far apart.

4.4.1 Explicit TV Eigenfunctions in 1D

Let us consider the function space setting and give an analytic expression of a large set of eigenfunctions of TV in one dimension with the signal domain being the entire real line. We will later see that Haar wavelets are a small subset of these, and hence eigenfunctions are expected to represent signals more concisely, with much fewer elements.

We give the presentation below in a somewhat informal manner with the goal of rather explaining the main concept of the findings than giving mathematical details. A formal presentation of the theory of TV eigenfunctions can be found, e.g., in [6].

The TV functional can be expressed as

$$J_{TV}(u) = \sup_{\substack{z \in C_c^\infty \\ \|z\|_{L^\infty(\Omega) \le 1}}} \langle u, \operatorname{div} z \rangle. \tag{4.9}$$

Let z_u be an argument admitting the supremum of (4.9); then, it immediately follows that $\operatorname{div} z_u \in \partial J_{TV}(u)$: in the one-homogeneous case we need to show $\langle u, \operatorname{div} z_u \rangle = J_{TV}(u)$ which is given in (4.9); and in addition that for any v in the space we have

$J_{TV}(v) \geq \langle v, p \rangle, p \in \partial J(u),$

$$J_{TV}(v) = \sup_{\|z\|_{L^\infty(\Omega)} \leq 1} \langle v, \operatorname{div} z \rangle \geq \langle v, \operatorname{div} z_u \rangle.$$

From here on, we will refer to z_u simply as z.

To understand better what z stands for, we can check the case of smooth u and perform integration by parts in (4.9) to have

$$J_{TV}(u) = \langle \nabla u, -z \rangle.$$

Then, as z also maximizes $\langle \nabla u, -z \rangle$, we can solve this pointwise, taking into account that $|z(x)| \leq 1$ and that the inner product of a vector is maximized for a vector at the same angle, to have

$$z(x) \begin{cases} = -\frac{\nabla u(x)}{|\nabla u(x)|} & \text{for } \nabla u(x) \neq \mathbf{0}, \\ \in [-1, 1] & \nabla u(x) = \mathbf{0}. \end{cases} \tag{4.10}$$

4.4.1.1 Single Peak

Let us consider the case of one-dimensional TV regularization and define the following function of a single unit peak of width w:

$$B_w(x) = \begin{cases} 1 & \text{for } x \in [0, w), \\ 0 & \text{otherwise.} \end{cases} \tag{4.11}$$

Then, it is well known (e.g., [3]) that any function of the type $h \cdot B_w(x - x_0)$ is an eigenfunction of TV in $x \in \mathbb{R}$ with eigenvalue $\lambda = \frac{2}{hw}$. Let us illustrate the latter by considering the characterization of the subdifferential (4.10) and define

$$z(x) = \begin{cases} -1 & \text{for } x \in]-\infty, x_0], \\ \frac{2(x - x_0)}{w} - 1 & \text{for } x \in (x_0, x_0 + w), \\ 1 & \text{for } x \in [x_0 + w, \infty[. \end{cases}$$

Fig. 4.3 A classical single-peak example of a TV eigenfunction in $\mathbb{R}$, $u(x) = B_w(x)$

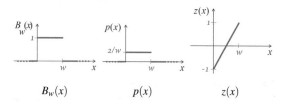

$B_w(x)$ $p(x)$ $z(x)$

Although this z is clearly not a C_c^∞ function, it was shown in [4, 6] that functions $z \in L^\infty(\mathbb{R})$ with div $z \in L^2(\mathbb{R})$ are sufficient for the characterization of subgradients.

For the above z, on the one hand, we have $\partial_x z(x) = \frac{w}{2} B_w(x - x_0)$, and on the other hand z meets Eq. (4.10) for $u = \frac{w}{2} B_w(\cdot - x_0)$. Therefore, $\frac{w}{2} B_w(\cdot - x_0) \in \partial J_{TV}(\frac{w}{2} B_w(\cdot - x_0))$, and after normalization with $\|\frac{w}{2} B_w(\cdot - x_0)\|_2 = \frac{w^2}{2}$ we find that $\frac{1}{w} B_w(\cdot - x_0)$ is an eigenfunction with eigenvalue $\lambda = \frac{2}{w}$. See Fig. 4.3 for an illustration of B_w, p and z.

4.4.1.2 Set of 1D Eigenfunctions

Generalizing this analysis, one can construct for any eigenvalue λ an infinite set of piecewise constant eigenfunctions (with a compact support).

Proposition 4.1 *Let* $-\infty < x_0 < x_1, .. < x_n < \infty$ *be a set of* $n + 1$ *points on the real line. Let*

$$u(x) = \sum_{i=0}^{n-1} h_i B_{w_i}(x - x_i), \qquad (4.12)$$

with $B_w(\cdot)$ *defined in (4.11),* $w_i = x_{i+1} - x_i$, *and*

$$h_i = \frac{2(-1)^i}{\lambda w_i}. \qquad (4.13)$$

Then, $u(x)$ *admits the eigenvalue problem (4.3) with* $J = J_{TV}$.

Proof One can construct the following z in the shape of "zigzag" between -1 and 1 at points x_i,

$$z(x) = (-1)^i \left(\frac{2(x - x_i)}{w_i} - 1 \right), \quad x \in [x_i, x_{i+1}),$$

and $\partial_x z = 0$ otherwise. In a similar manner to the single-peak case, we get the subgradient element in $\partial J_{TV}(u)$

$$p(x) = \begin{cases} \partial_x z = (-1)^i \frac{2}{w_i}, & x \in [x_i, x_{i+1}) \\ 0, & x \notin [x_0, x_n). \end{cases} \qquad (4.14)$$

This yields $p(x) = \lambda u(x)$.

In Figs. 4.4 and 4.5, we see examples of u, p, and z which construct simple TV eigenfunctions in the unbounded and bounded domains, respectively. In Fig. 4.6, several different types of eigenfunctions are shown. Compared to wavelets, they also oscillate (with mean zero value) but their variability is significantly larger.

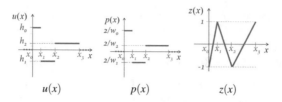

u(x)   p(x)   z(x)

Fig. 4.4 Illustration of Proposition 4.1, a TV eigenfunction $u(x)$ in $\mathbb{R}$

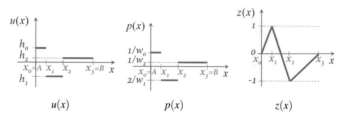

u(x)   p(x)   z(x)

Fig. 4.5 A TV eigenfunction $u(x)$ in a bounded domain

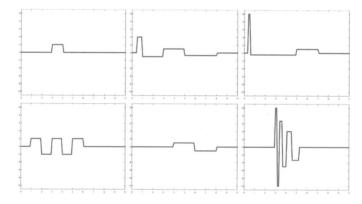

Fig. 4.6 A few examples of functions meeting $f \in \partial J_{TV}(f)$ in $\mathbb{R}$. (Note that these are piecewise constant functions and the jumps should be understood as discontinuities)

4.4.1.3 The Bounded Domain Case

The unbounded domain is easier to analyze in some cases; however, in practice we can implement only signals with a bounded domain. We therefore give the formulation for $u \in \Omega = [A, B] \subset \mathbb{R}$. Requiring $J_{TV}(1) = \langle\, p, 1\rangle = 0$ and $\langle u, p\rangle = \langle -\nabla u, z\rangle$ leads to the boundary condition $z|_{\partial\Omega} = 0$.

Thus, on the boundaries, we have $z = 0$ with half the slope of the unbounded case (see Fig. 4.5), all other derivations are the same. Setting $x_0 = A$, $x_n = B$, we get the solution of (4.12) with h_i defined slightly differently as

$$h_i = \frac{2a_i(-1)^i}{\lambda w_i}, \tag{4.15}$$

where $a_i = \frac{1}{2}$ for $i \in \{1, n-1\}$ and $a_i = 1$ otherwise.

Remark: Note that if u is an eigenfunction so is $-u$ so the formulas above are all valid also with the opposite sign.

4.5 Pseudo-Eigenfunctions

The first introduction to the idea of *pseudospectra* was given by Landau [26], who used the term ε spectrum. Further, extension of the topic was given in [27, 28], generalizing the theory for matrices and linear operators. Trefethen coined the term *pseudospectra* [29, 30] presenting an overview of the theory and applications in [31].

Given two linear operators L and E, a pseudo-eigenfunction u of L admits the following eigenvalue problem:

$$(L+E)u = \lambda u, \quad \text{s.t. } \|E\| \le \varepsilon. \tag{4.16}$$

That is, u is an eigenfunction of an operator which is very similar to L, up to a small perturbation. The corresponding value λ is said to be a *pseudo-eigenvalue*, or more precisely an element in the ε-*pseudosepctra* of L. Note that λ does not have to be close to any eigenvalue of L, but is an exact eigenvalue of some operator similar to L.

For nonlinear operators, it is not trivial how this notion could be generalized (as two operators cannot simply be added). Therefore, we define a somewhat different notion, which we refer to as a *measure of affinity to eigenfunctions*. The measure is in the range $[0, 1]$ and attains a value of 1 for eigenfunctions (and only for them). When it is very close to 1, this can be considered as an alternative definition of a pseudo-eigenfunction, which is a very useful notion in the discrete and graph case, as one may not be able to obtain a precise nonlinear eigenfunction in all cases (but may reach numerically a good approximation). We show below the exact relation for the linear case.

4.5.1 Measure of Affinity of Nonlinear Eigenfunctions

Let T be a general nonlinear operator in a Banach space $\mathscr{X}$, $T : \mathscr{X} \to \mathscr{X}$ embedded in L^2 such that $T(u) \in L^2$. The corresponding nonlinear eigenvalue problem is (4.2), $(T(u) = \lambda u)$.

Definition 4.1 The measure $\mathscr{A}_T(u)$ of the affinity of a function u to an eigenfunction, based on the operator T, with $\|u\| \neq 0$, $\|T(u)\| \neq 0$, is defined by

$$\mathscr{A}_T(u) := \frac{|\langle u, T(u)\rangle|}{\|u\| \cdot \|T(u)\|}. \tag{4.17}$$

Proposition 4.2 $\mathscr{A}_T(u)$ admits the following:

$$0 \leq \mathscr{A}_T(u) \leq 1, \quad \mathscr{A}_T(u) = 1 \ \textit{iff } u \textit{ admits the eigenvalue problem}. \tag{4.18}$$

Proof This is an immediate consequence of the Cauchy–Schwarz inequality.

That is, the measure is 1 for all eigenfunctions and only for them (we remind that for the Cauchy–Schwarz inequality equality is attained if and only if the two functions are linearly dependent). The measure then has a graceful degradation from 1 to 0.

Let us define the projection of u onto the plane orthogonal to $T(u)$:

$$w := u - \frac{\langle u, T(u)\rangle}{\|T(u)\|^2} T(u).$$

Then $\mathscr{A}_T(u)$ decreases as $\|w\|$ increases, where for eigenfunctions $\|w\| = 0$. Using the above, we determine a pseudo-eigenfunction being close up to ε to an exact eigenfunction of a nonlinear operator, if the following bound on $\mathscr{A}_T(u)$ holds:

$$\mathscr{A}_T(u) \geq 1 - \varepsilon. \tag{4.19}$$

Geometric interpretation of the measure. Considering Definition 4.1, it can be written as $\mathscr{A}_T(u) = \cos(\theta)$, i.e., $\mathscr{A}_T(u)$ is based on the angle between u and $T(u)$. Thus, it may be more insightful to look at θ itself,

$$\theta = \cos^{-1}(\mathscr{A}_T(u)). \tag{4.20}$$

Fig. 4.7 An illustration of the angle induced by u and $T(u)$ (an operator acting on u). Figure **a** shows the case when u is an arbitrary function, while figure **b** illustrates the case when u is an eigenfunction of T

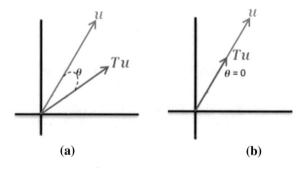

(a)   (b)

An illustration of two cases, non-eigenfunction (a) and eigenfunction (b), is shown in Fig. 4.7.

4.5.1.1   The One-Homogeneous and Linear Case

For eigenfunctions induced by convex one-homogeneous functionals, we give the following adaptation of (4.17):

$$A_{p(u)}(u) = \frac{J(u)}{\|p(u)\| \cdot \|u\|} = \frac{\langle p(u), u \rangle}{\|p(u)\| \cdot \|u\|}, \tag{4.21}$$

for all $p(u) \in \partial J(u)$ (and having $J(u) \geq 0$ the absolute expression in the numerator can be omitted).

Having a linear operator L, the definition is now given by

$$A_L(u) = \frac{\langle Lu, u \rangle}{\|Lu\| \cdot \|u\|}. \tag{4.22}$$

We would like to show a connection between $A_L(u)$ and the pseudo-eigenfunction definition given in (4.16). For simplicity, without loss of generality, we rescale the norm such that the norm of the identity is one, i.e., $\|I\| = 1$ (as naturally, norms are equivalent under a constant multiplication). Let u admit (4.16), then we can rewrite (4.16) as

$$Lu = \lambda u - Eu,$$

and we have

$$\|Lu\| = \|\lambda u - Eu\| = \|(\lambda I - E)u\| \leq \|\lambda I - E\| \cdot \|u\| \leq (\|\lambda I\| + \|E\|)\|u\|$$
$$= (\lambda + \|E\|)\|u\| \leq (\lambda + \varepsilon)\|u\|.$$

Using (4.18) and (4.16), we can write

$$\begin{aligned}
1 = A_{L+E}(u) &= \frac{\langle (L+E)u, u \rangle}{\|(L+E)u\| \cdot \|u\|} = \frac{\langle Lu, u \rangle}{\|(L+E)u\| \cdot \|u\|} + \frac{\langle Eu, u \rangle}{\|(L+E)u\| \cdot \|u\|} \\
&= \frac{\langle Lu, u \rangle \cdot \|Lu\|}{\|(L+E)u\| \cdot \|u\| \cdot \|Lu\|} + \frac{\langle Eu, u \rangle}{\|(L+E)u\| \cdot \|u\|} \\
&= \frac{A_L(u)\|Lu\|}{\|\lambda u\|} + \frac{\langle Eu, u \rangle}{\|\lambda u\| \cdot \|u\|},
\end{aligned}$$

where we use $(L + E)u = \lambda u$. Using the result, we have for $\|Lu\|$ and the Cauchy–Schwarz inequality for the second expression we get

$$
\begin{aligned}
&\leq A_L(u)\frac{(\lambda + \varepsilon)\|u\|}{\|\lambda u\|} + \frac{\|Eu\| \cdot \|u\|}{\|\lambda u\| \cdot \|u\|} \leq A_L\frac{(\lambda + \varepsilon)\|u\|}{\|\lambda u\|} + \frac{\|E\| \cdot \|u\|}{\|\lambda u\|} \\
&= A_L\frac{\lambda + \varepsilon}{\lambda} + \frac{\|E\|}{\lambda} \\
&\leq A_L\frac{\lambda + \varepsilon}{\lambda} + \frac{\varepsilon}{\lambda}.
\end{aligned}
$$

We can conclude that

$$
A_L(u) \geq \frac{\lambda - \varepsilon}{\lambda + \varepsilon}. \tag{4.23}
$$

References

1. J. Appell, E. De Pascale, A. Vignoli, *Nonlinear Spectral Theory*, vol. 10 (Walter de Gruyter, 2004)
2. L. Rudin, S. Osher, E. Fatemi, Nonlinear total variation based noise removal algorithms. Physica D **60**, 259–268 (1992)
3. Y. Meyer, Oscillating patterns in image processing and in some nonlinear evolution equations, March 2001. The 15th Dean Jacquelines B. Lewis Memorial Lectures
4. F. Andreu, C. Ballester, V. Caselles, J.M. Mazón, Minimizing total variation flow. Differ. Integral Equ. **14**(3), 321–360 (2001)
5. F. Andreu, V. Caselles, J.I. Dıaz, J.M. Mazón, Some qualitative properties for the total variation flow. J. Funct. Anal. **188**(2), 516–547 (2002)
6. G. Bellettini, V. Caselles, M. Novaga, The total variation flow in R^N. J. Differ. Equ. **184**(2), 475–525 (2002)
7. G. Steidl, J. Weickert, T. Brox, P. Mrzek, M. Welk, On the equivalence of soft wavelet shrinkage, total variation diffusion, total variation regularization, and SIDEs. SIAM J. Numer. Anal. **42**(2), 686–713 (2004)
8. M. Burger, K. Frick, S. Osher, O. Scherzer, Inverse total variation flow. Multiscale Model. Simul. **6**(2), 366–395 (2007)
9. S. Bartels, R.H. Nochetto, J. Abner, A.J. Salgado, *Discrete Total Variation Flows Without Regularization* (2012), arXiv:1212.1137
10. Y. Giga, R.V. Kohn, Scale-invariant extinction time estimates for some singular diffusion e-quations. Hokkaido Univ. Preprint Ser. Math. (963) (2010)
11. M. Burger, G. Gilboa, S. Osher, J. Xu, Nonlinear inverse scale space methods. Commun. Math. Sci. **4**(1), 179–212 (2006)
12. J. Müller, Advanced image reconstruction and denoising: Bregmanized (higher order) total variation and application in pet, 2013. Ph.D. Thesis, Univ. Münster
13. M. Benning, C. Brune, M. Burger, J. Müller, Higher-order tv methods: enhancement via bregman iteration. J. Sci. Comput. **54**, 269–310 (2013)
14. C. Pöschl, O. Scherzer, *Exact Solutions of One-dimensional TGV* (2013), arXiv:1309.7152, 2013
15. K. Papafitsoros, K. Bredies, A Study of the One Dimensional Total Generalised Variation Regularisation Problem (2013), arXiv:1309.5900
16. K. Bredies, K. Kunisch, T. Pock, Total generalized variation. SIAM J. Imag. Sci. **3**(3), 492–526 (2010)

17. A. Chambolle, P.L. Lions, Image recovery via total variation minimization and related problems. Numerische Mathematik **76**(3), 167–188 (1997)
18. J. Duran, M. Moeller, C. Sbert, D. Cremers, *Collaborative Total Variation: A General Framework for Vectorial TV Models*. Submitted, arXiv:1508.01308
19. M. Welk, G. Steidl, J. Weickert, Locally analytic schemes: a link between diffusion filtering and wavelet shrinkage. Appl. Comput. Harmon. Anal. **24**(2), 195–224 (2008)
20. M. Benning, M. Burger, Ground states and singular vectors of convex variational regularization methods. Methods Appl. Anal. **20**(4), 295–334 (2013)
21. L. Dorst, R. Van den Boomgaard, Morphological signal processing and the slope transform. Signal Process. **38**(1), 79–98 (1994)
22. U. Köthe, Local appropriate scale in morphological scale-space, in *ECCV'96* (Springer, Berlin, 1996), pp. 219–228
23. M. Burger, G. Gilboa, M. Moeller, L. Eckardt, D. Cremers, Spectral decompositions using one-homogeneous functionals. SIAM J. Imag. Sci. **9**(3), 1374–1408 (2016)
24. K. Dabov, A. Foi, V. Katkovnik, K. Egiazarian, Image denoising by sparse 3-d transform-domain collaborative filtering. IEEE Trans. Image Process. **16**(8), 2080–2095 (2007)
25. D. Zoran, Y. Weiss, From learning models of natural image patches to whole image restoration, in *2011 International Conference on Computer Vision* (IEEE, 2011), pp. 479–486
26. H.J. Landau, On szegö's eingenvalue distribution theorem and non-hermitian kernels. Journal d'Analyse Mathématique **28**(1), 335–357 (1975)
27. J.M. Varah, On the separation of two matrices. SIAM J. Numer. Anal. **16**(2), 216–222 (1979)
28. F. Chatelin, The spectral approximation of linear operators with applications to the computation of eigenelements of differential and integral operators. SIAM Rev. **23**(4), 495–522 (1981)
29. L.N. Trefethen, Approximation theory and numerical linear algebra, in *Algorithms for Approximation II* (Springer, Berlin, 1990), pp. 336–360
30. N. Lloyd, Trefethen. Pseudospectra of matrices. Numer. Anal. **91**, 234–266 (1991)
31. L.N. Trefethen, M. Embree, *Spectra and Pseudospectra: the Behavior of Nonnormal Matrices and Operators* (Princeton University Press, Princeton, 2005)

Chapter 5
Spectral One-Homogeneous Framework

5.1 Preliminary Definitions and Settings

We will consider convex functionals J either in a function space setting $J : \mathscr{X} \to \mathbb{R}$ for a Banach space $\mathscr{X}$ embedded into L^2 or—in a discrete setting—as a function $J : \mathbb{R}^n \to \mathbb{R}$. We will specify the considered setting at those points where it is crucial.

5.2 Spectral Representations

5.2.1 Scale Space Representation

We will use the scale space evolution, which is straightforward, as the canonical case of spectral representation. Consider a convex (absolutely) one-homogeneous functional J.

The *scale space* or *gradient flow* is

$$\partial_t u(t) = -p(t), \quad p(t) \in \partial J(u(t)), \; u(0) = f. \tag{5.1}$$

Linear and nonlinear scale space techniques have been used for various image processing algorithms, as outlined in Chap. 2.

A spectral representation based on the gradient flow formulation (5.1) was the first work toward defining a nonlinear spectral decomposition and has been conducted in [1, 2] for the case of J being the TV regularization. In [3], the author and colleagues extended this notion to general one-homogeneous functionals by observing that the solution of the gradient flow can be computed explicitly for any one-homogeneous J in the case of f being an eigenfunction. For $\lambda f \in \partial J(f)$, the solution to (5.1) is given by

© Springer International Publishing AG, part of Springer Nature 2018
G. Gilboa, *Nonlinear Eigenproblems in Image Processing and Computer Vision*, Advances in Computer Vision and Pattern Recognition,
https://doi.org/10.1007/978-3-319-75847-3_5

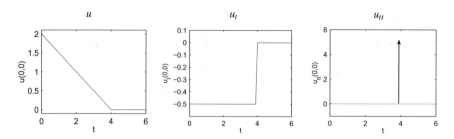

Fig. 5.1 Illustrating the TV-flow evolution of a disk in $\mathbb{R}^2$. The value is within $|x| < r$, for example, at $(x_1 = 0, x_2 = 0)$. The second derivative is an impulse at time t_d. [here, we set $r = 4, h = 2$, and therefore $t_d = 1/\lambda = rh/2 = 4$]

$$u(t) = \begin{cases} (1 - t\lambda)f & \text{for } t \le \frac{1}{\lambda}, \\ 0 & \text{else.} \end{cases} \tag{5.2}$$

Note that in linear spectral transformations such as Fourier or Wavelet-based approaches, the input data being an eigenfunction leads to the energy of the spectral representation being concentrated at a single wavelength. To preserve this behavior for nonlinear spectral representations, the wavelength decomposition of the input data f is defined by

$$\phi(t) = t\partial_{tt}u(t). \tag{5.3}$$

In Fig. 5.1, we show the basic intuition of using the second time derivative to isolate eigenfunctions. The main principle is that linearly diminishing structures, such as eigenfunctions in a one-homogeneous gradient flow, respond only at a singular flow-time. This can be viewed as separating all linearly diminishing structures (which, as we see later, are also difference of two eigenfunctions). Note that due to the piecewise linear behavior in (5.2), the wavelength representation of an eigenfunction f becomes $\phi(t) = \delta(t - \frac{1}{\lambda})f$, where δ denotes a Dirac delta distribution. The name *wavelength* decomposition is natural because for $\lambda f \in \partial J(f)$, $\| f \| = 1$, one readily shows that $\lambda = J(f)$, which means that the eigenvalue λ is corresponding to a generalized frequency. In analogy to the linear case, the inverse relation of a peak in ϕ appearing at $t = \frac{1}{\lambda}$ motivates the interpretation as a wavelength, as discussed in more details in the following section.

For arbitrary input data f, one can reconstruct the original image by

$$f(x) = \int_0^\infty \phi(t; x)dt. \tag{5.4}$$

Given a *transfer function* $H(t) \in \mathbb{R}$, image filtering can be performed by

$$f^H(x) := \int_0^\infty H(t)\phi(t; x)dt. \tag{5.5}$$

We would like to point out that Eq. (5.3) is an informal definition of the wavelength decomposition which is supposed to illustrate the general idea of nonlinear spectral decompositions. Below, a more formal presentation is given which gives sense to (5.3) for one-homogeneous J in the spatially discrete setting as elements in $(W_{loc}^{1,1}(\mathbb{R}^+, \mathbb{R}^n))^*$. The discrete case also permits to show a sufficiently rapid decrease of $\partial_t u(t)$ for the reconstruction (5.5) to hold for all absolutely one-homogeneous regularization functions J.

5.3 Signal Processing Analogy

Up until very recently, nonlinear filtering approaches such as (5.1) or related variational methods have been treated independently of the classical linear point of view of changing the representation of the input data, filtering the resulting representation and inverting the transform. In [1, 2], it was proposed to use (5.1) in the case of J being the total variation to define a TV spectral representation of images that allows to extend the idea of filtering approaches from the linear to the nonlinear case. This was later generalized to one-homogeneous functionals in [3].

Classical Fourier filtering has some very convenient properties for analyzing and processing signals:

1. Processing is performed in the transform (frequency) domain by simple attenuation or amplification of desired frequencies.
2. A straightforward way to visualize the frequency activity of a signal is through its spectrum plot. The spectral energy is preserved in the frequency domain, through the well-known Parseval's identity.
3. The spectral decomposition corresponds to the coefficients representing the input signal in a new orthonormal basis.
4. Both transform and inverse transform are linear operations.

In the nonlinear setting, the first two characteristics are mostly preserved. Orthogonality could so far only be obtained for the particular (discrete) case of $J(u) = \|Vu\|_1$ with VV^* being diagonally dominant, [4]. Further, orthogonality statements are still an open issue. Finally, linearity is certainly lost. In essence, we obtain a *nonlinear* forward transform and a *linear* inverse transform. Thus, following the nonlinear decomposition, filtering can be performed easily.

In addition, we gain edge preservation and new scale features, which, unlike sines and cosines, are data-driven and are therefore highly adapted to the image. Thus, the filtering has far less tendency to create oscillations and artifacts.

Let us first derive the relation between Fourier and the eigenvalue problem (4.3). For

$$J(u) = \frac{1}{2} \int |\nabla u(x)|^2 dx,$$

we get $-\Delta u \in \partial J(u)$. Thus, with appropriate boundary conditions sines and cosine, the basis functions of the Fourier transform are solutions of this eigenvalue problem.

Fig. 5.2 Comparison
between the ideal low-pass
filter response and TV flow.
In both cases, the response is
shown for the minimal extent
of filtering in which the
smallest circle completely
vanishes. One sees the
considerable reduction of
contrast of the larger circles
in the TV flow versus the
sharp and stable results of
the ideal TV LPF

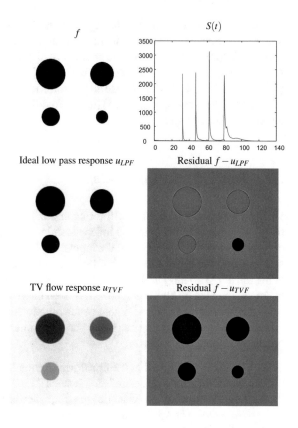

In this case, for a frequency ω of the sine function, $\sin(\omega x)$, we have the relation
$\lambda = \omega^2$. Other convex regularizing functionals, such as TV and TGV, can therefore
be viewed as natural nonlinear generalizations.

A fundamental concept in linear filtering is *ideal filters* [5]. Such filters either
retain or diminish completely frequencies within some range. In a linear time (space)
invariant system, a filter is fully determined by its transfer function $H(\omega)$. The filtered
response of a signal $f(x)$, with Fourier transform $F(\omega)$, filtered by a filter H is

$$f^H(x) := \mathscr{F}^{-1}(H(\omega) \cdot F(\omega)),$$

with $\mathscr{F}^{-1}$ the inverse Fourier transform. For example, an ideal low-pass filter retains
all frequencies up to some cutoff frequency ω_c. Its transfer function is thus

$$H(\omega) = \begin{cases} 1 & \text{for } 0 \le \omega \le \omega_c, \\ 0 & \text{else.} \end{cases}$$

Viewing frequencies as eigenvalues in the sense of (4.3), one can define generalizations of these notions.

5.3.1 Nonlinear Ideal Filters

The (ideal) low-pass filter (LPF) can be defined by Eq. (5.5) with $H(t) = 1$ for $t \geq t_c$ and 0 otherwise, or

$$LPF_{t_c}(f) := \int_{t_c}^{\infty} \phi(t; x)dt. \tag{5.6}$$

Its complement, the (ideal) high-pass filter (HPF), is defined by

$$HPF_{t_c}(f) := \int_{0}^{t_c} \phi(t; x)dt. \tag{5.7}$$

Similarly, band (pass/stop) filters are filters with low and high cutoff scale parameters $(t_1 < t_2)$

$$BPF_{t_1,t_2}(f) := \int_{t_1}^{t_2} \phi(t; x)dt, \tag{5.8}$$

$$BSF_{t_1,t_2}(f) := \int_{0}^{t_1} \phi(t; x)dt + \int_{t_2}^{\infty} \phi(t; x)dt. \tag{5.9}$$

For f being a single eigenfunction with eigenvalue $\lambda_0 = \frac{1}{t_0}$, the above definitions coincide with the linear definitions, where the eigenfunction is either completely preserved or completely diminished, depending on the cutoff eigenvalue(s) of the filter. As an example see Fig. 5.2 where an ideal spectral low-pass filter which aims at removing the smallest disk is compared to a TV flow attempting the same task. The ideal filter is performing this separation task in almost a perfect manner, without contrast loss of the other scales (larger disks).

Again the above point of view is rather illustrative than mathematically precise. As mentioned above, the discrete case allows to identify ϕ with elements of $(W_{loc}^{1,1})^*$ such that only mollified version of the above ideal filters are permissible, see [4] for details.

In Fig. 5.3, an example of spectral TV processing is shown with the response of the four filters defined above in Eqs. (5.6)–(5.9).

5.3.2 Spectral Response

As in the linear case, it is very useful to measure in some sense the "activity" at each frequency (scale). This can help identify dominant scales and design better the filtering strategies (either manually or automatically). Moreover, one can obtain a notion of the type of energy which is preserved in the new representation using some analog of Parseval's identity.

In [1, 2], a L^1 type spectrum was suggested for the TV spectral framework (without trying to relate to a Parseval rule),

Fig. 5.3 Total variation
spectral filtering example.
The input image (top left) is
decomposed into its $\phi(t)$
components, the
corresponding spectrum
$S_1(t)$, Eq. (5.10), is on the
top right. Integration of the
ϕ's over the t domains 1, 2,
and 3 (top right) yields
high-pass, band-pass, and
low-pass filters, respectively
(chosen manually here). The
band-stop filter (bottom
right) is the complement
integration domain of region
2. Taken from [3]

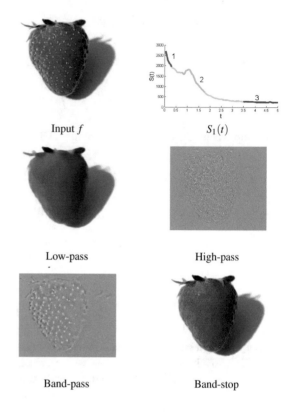

Input f $S_1(t)$

Low-pass   High-pass

Band-pass   Band-stop

$$S_1(t) := \|\phi(t;x)\|_{L^1(\Omega)} = \int_\Omega |\phi(t;x)| dx. \qquad (5.10)$$

In [3], the following definition was suggested:

$$S_2^2(t) = t^2 \frac{d^2}{dt^2} J(u(t)) = \langle \phi(t), 2tp(t) \rangle. \qquad (5.11)$$

With this definition, the following analog of the Parseval identity was shown:

$$\|f\|^2 = -\int_0^\infty \frac{d}{dt} \|u(t)\|^2 \, dt$$

$$= 2 \int_0^\infty \langle p(t), u(t) \rangle \, dt$$

$$= 2 \int_0^\infty J(u(t)) \, dt$$

$$= \int_0^\infty S_2^2(t) \, dt. \qquad (5.12)$$

In [4], a third definition for the spectrum was suggested (which is simpler and admits Parseval):

$$S_3^2(t) = \langle \phi(t), f \rangle. \tag{5.13}$$

It can be shown that a similar Parseval-type equality holds here

$$\|f\|^2 = \langle f, f \rangle = \langle \int_0^\infty \phi(t)dt, f \rangle = \int_0^\infty \langle \phi(t), f \rangle dt$$

$$= \int_0^\infty S_3^2(t) \, dt. \tag{5.14}$$

While (5.10), (5.11), and (5.13) yield the intuition of possible definitions of the spectral decomposition, care has to be taken to obtain mathematically precise definitions. In the discrete case, we have $\phi \in (W_{loc}^{1,1}(\mathbb{R}^+, \mathbb{R}^n))^*$, such that S_2^2 and S_3^2 have to

Table 5.1 Some properties of the TV transform in $\mathbb{R}^n$, compared to the Fourier transform. The symbol $+$ denotes that the property holds for general one-homogeneous functionals

	TV transform	Fourier transform
Transform	Gradient flow:$+$ $\phi(t; x) = t\partial_{tt}u, u_t \in -\partial J(u).$ $u\vert_{t=0} = f$	$F(\omega) = \int f(x)e^{-i\omega x}dx$
Inverse transform	$f(x) = \int_0^\infty \phi(t; x)dt + \bar{f}$ $+$	$f(x) = \frac{1}{(2\pi)^n}\int F(\omega)e^{i\omega x}d\omega$
Eigenfunctions	$\lambda u \in \partial J(u)$ $+$	$e^{i\omega x}$
Spectrum—amplitude	Three alternatives: $+$ $S_1(t) = \|\phi(t; x)\|_{L^1},$ $S_2(t)^2 = \langle \phi(t), 2tp(t) \rangle,$ $S_3(t)^2 = \langle \phi(t), f \rangle$	$\vert F(\omega) \vert$
Translation $f(x - a)$	$\phi(t; x - a)$	$F(\omega)e^{-ia\omega}$
Rotation $f(Rx)$	$\phi(t; Rx),$	$F(R\omega)$
Contrast change $af(x)$	$\phi(t/a; x)$	$aF(\omega)$
Spatial scaling $f(ax)$	$a\phi(at; ax)$	$\frac{1}{\vert a \vert}F(\frac{\omega}{a})$
Linearity $f_1(x) + f_2(x)$	Not in general	$F_1(\omega) + F_2(\omega)$
Parseval's identity	$\int \vert f(x) \vert^2 dx = \int S_2(t)^2 dt =$ $\int S_3(t)^2 dt$ $+$	$\int \vert f(x) \vert^2 dx = \int \vert F(\xi) \vert^2 d\xi,$ $\xi := \frac{\omega}{2\pi}$
Orthogonality	$\langle \phi(t), u(t) \rangle = 0$ $+$	$\langle e^{i\omega_1 x}, e^{i\omega_2 x} \rangle = 0, \omega_1 \neq \omega_2$
Open issues (some)		
Fourier duality	–	$F(x) \leftrightarrow f(-\xi), \xi := \frac{\omega}{2\pi}$
Convolution	–	$f_1(x) * f_2(x) \leftrightarrow F_1(\omega)F_2(\omega)$
Spectrum—phase	–	$\tan^{-1}(\Im(F(\omega))/\Re(F(\omega)))$

be interpreted as elements in $(W^{1,1}_{loc}(\mathbb{R}^+, \mathbb{R}))^*$. In fact, in [4], we showed these two definitions to be equivalent for sufficiently regular subdifferentials of J.

As an overview example, Table 5.1 summarizes the analogies of the Fourier and total variation based spectral transformation. There remain many open questions of further possible generalizations regarding Fourier-specific results, some of which are listed like the convolution theorem, Fourier duality property (where the transform and inverse transform can be interchanged up to a sign), the existence of phase in Fourier, and more.

5.4   Theoretical Analysis and Properties

We now address some of the above concepts more formally. We can define nonlinear filtering of the data f with respect to J via certain integrals with respect to the measure $\tilde{\Phi}_s$. If we are merely interested in the latter with sufficiently regular filters, we can extend from $\tilde{\Phi}_s$ to vector-valued distributions.

Definition 5.1 (*Weak spectral representation*) A map from $f \in \mathscr{X}$ to a vector-valued distribution $\tilde{\phi}(s)$ on $\mathscr{X}$ is called a *weak spectral representation with respect to the convex functional J* if the following properties are satisfied:

- **Eigenvectors as atoms**: For f satisfying $\|f\| = 1$ and $\lambda f \in \partial J(f)$, the weak spectral representation is given by $\tilde{\phi}(s) = f \, \delta_\lambda(s)$.
- **Reconstruction**: The input data f can be reconstructed by

$$f = \int_0^\infty \tilde{\phi}(s) \, ds. \tag{5.15}$$

Again, exactly the same definition of a weak spectral representation can also be made for spectral wavelength decomposition with an inverse relation between λ and t for eigenvectors as atoms.

We mention that throughout this section we will use a rather formal notation for the distribution $\tilde{\phi}$ and its wavelength counterpart ϕ as in the reconstruction formula above. All integrals we write indeed have to be understood as duality products with sufficiently smooth test functions. We will verify the well-definedness for the specific spectral representations we investigate below; it will turn out that in all cases it suffices to use test functions in $W^{1,1}_{loc}(\mathbb{R}_+; \mathbb{R}^n)$.

5.4.1   Variational Representation

In this section, we would like to describe how to define a spectral representation based on the variational method. The standard regularization with L^2 square fidelity is defined by

$$u_{VM}(t) = \arg\min_u \frac{1}{2}\|u - f\|_2^2 + tJ(u), \tag{5.16}$$

for a suitable proper, convex, lower semi-continuous regularization functional J : $\mathcal{X} \to \mathbb{R}^+ \cup \{\infty\}$ defined on Banach space $\mathcal{X}$ embedded into $L^2(\Omega)$.

We would like to establish the following analogy to the linear spectral analysis: Eigenfunctions meeting (4.3) should be fundamental atoms of the spectral decomposition. It is easy to verify (cf. [6]) that for f being an eigenfunction the solution to (5.16) is given by

$$u_{VM}(t) = \begin{cases} (1 - t\lambda)f & \text{for } t \le \frac{1}{\lambda}, \\ 0 & \text{else.} \end{cases}$$

Since $u_{VM}(t)$ behaves piecewise linear in time, and our goal is to obtain a single peak for f being an eigenfunction, it is natural to consider the second derivative of $u_{VM}(t)$ (in a distributional sense). The latter will yield the (desired) delta distribution $\partial_{tt} u_{VM}(t) = \lambda \delta_{1/\lambda}(t)f$. Consider the desired property of reconstructing the data by integrating the spectral decomposition. We find that in the case of f being an eigenfunction

$$\int_0^\infty \partial_{tt} u_{VM}(t)\, dt = \lambda f.$$

Therefore, a normalization by multiplying $\partial_{tt} u_{VM}(t)$ with t is required to fully meet the integration criterion. Motivated by the behavior of the variational method on eigenfunctions, we make the following definition:

Definition 5.2 (*Spectral Representation based on* (5.16)) Let $u_{VM}(t)$ be the solution to (5.16). We define

$$\phi_{VM}(t) = t\partial_{tt} u_{VM}(t) \tag{5.17}$$

to be the *wavelength decomposition* of f.

In the following, we will show that the above ϕ_{VM} is a weak spectral wavelength decomposition in the sense of Definition 5.1 for any convex absolutely one-homogeneous regularization functional J.

Let us explain the term wavelength. In the case of f being an eigenvector, we can see that the peak in $\phi_{VM}(t)$ appears at a later time, the smaller λ is. The eigenvalue λ reflects our understanding of generalized frequencies which is nicely underlined by the fact that $\lambda = J(f)$ holds due to the absolute one-homogenity of J. Therefore, we expect contributions of small frequencies at large t and contributions of high frequencies at small t, which is the relation typically called wavelength representation in the linear setting.

While we have seen that the definition of (5.17) makes sense in the case of f being an eigenfunction, it remains to show this for general f. As a first step, we state the following property of variational methods with absolutely one-homogeneous regularizations:

Proposition 5.1 (Finite time extinction) *Let J be an absolutely one-homogeneous functional, and f be arbitrary. Then, there exists a time $T < \infty$ such that u_{VM} determined by (5.16) meets*

$$u_{VM}(T) = P_0(f).$$

Proof Considering the optimality conditions for (5.16), the above statement is the same as

$$\frac{Q_0 f}{T} = \frac{f - P_0(f)}{T} \in \partial J(P_0(f)) = \partial J(0).$$

Since $\partial J(0)$ has nonempty relative interior in $\mathcal{N}(J)$, this is guaranteed for T sufficiently large.

Note that the above proof yields the extinction as the minimal value T such that $\frac{Q_0 f}{T} \in \partial J(0)$, which can also be expressed as the minimal T such that $J^*(\frac{Q_0 f}{T}) = 0$, i.e., $\frac{Q_0 f}{T}$ is in the dual unit ball. This is the generalization of the well-known result by Meyer [7] for total variation denoising.

A second useful property is the Lipschitz continuity of u_{VM}, which allows to narrow the class of distributions for ϕ:

Proposition 5.2 *The function $u_{VM} : \mathbb{R}^+ \to \mathbb{R}^n$ is Lipschitz continuous. Moreover, the spectral representation satisfies $\phi_{VM} \in (W_{loc}^{1,1}(\mathbb{R}^+, \mathbb{R}^n))^*$.*

Proof Consider $u_{VM}(t)$ and $u_{VM}(t + \Delta t)$. Subtracting the optimality conditions yields

$$0 = u_{VM}(t) - u_{VM}(t + \Delta t) + t(p_{VM}(t) - p_{VM}(t + \Delta t)) - \Delta t p_{VM}(t + \Delta t).$$

Taking the inner product with $u_{VM}(t) - u_{VM}(t + \Delta t)$ yields

$$0 = \|u_{VM}(t) - u_{VM}(t + \Delta t)\|^2 - \Delta t \langle p_{VM}(t + \Delta t), u_{VM}(t) - u_{VM}(t + \Delta t) \rangle$$
$$+ t \underbrace{\langle p_{VM}(t) - p_{VM}(t + \Delta t)), u_{VM}(t) - u_{VM}(t + \Delta t) \rangle}_{\geq 0}$$
$$\geq \|u_{VM}(t) - u_{VM}(t + \Delta t)\|^2 - \Delta t \langle p_{VM}(t + \Delta t), u_{VM}(t) - u_{VM}(t + \Delta t) \rangle$$
$$\geq \|u_{VM}(t) - u_{VM}(t + \Delta t)\|^2 - \Delta t \|p_{VM}(t + \Delta t)\| \|u_{VM}(t) - u_{VM}(t + \Delta t)\|.$$

Using Conclusion 1 of Chap. 1 (and Lemma 1.5), we find

$$\|u_{VM}(t) - u_{VM}(t + \Delta t)\| \leq \Delta t \, C.$$

Through integration by parts, we obtain for regular test functions v:

$$\int_0^\infty v(t) \cdot \phi_{VM}(t) \, dt = \int_0^\infty t v(t) \cdot \partial_{tt} u_{VM}(t) \, dt,$$

$$= -\int_0^\infty \partial_t (v(t) \, t) \partial_t u_{VM}(t) \, dt,$$

$$= -\int_0^\infty (t \partial_t v(t) + v(t)) \, \partial_t u_{VM}(t) \, dt. \tag{5.18}$$

If $u_{VM}(t)$ is Lipschitz continuous, then $\partial_t u_{VM}(t)$ is an L^∞ function. Due to the finite time extinction, the above integrals can be restricted to the interval $(0, T_{ext})$ and the last integral is well defined for any v such that $v \in W_{loc}^{1,1}(\mathbb{R}_+, \mathbb{R}^n)$. A standard density argument yields that (5.21) finally allows to use all such test functions, i.e., defines $\partial_t u_{VM}$ in the dual space.

Thanks to the finite time extinction, we can state the reconstruction of any type of input data f by integration over $\phi_{VM}(t)$ in general.

Theorem 5.1 (Reconstruction of the input data) *It holds that*

$$f = P_0(f) + \int_0^\infty \phi_{VM}(t) \, dt. \tag{5.19}$$

Proof Since for each vector $g \in \mathbb{R}^n$ the constant function $v = g$ is an element of $W_{loc}^{1,1}(\mathbb{R}_+, \mathbb{R}^n)$, we can use (5.18)

$$\int_0^\infty g \cdot \phi_{VM}(t) \, dt = -\int_0^\infty g \cdot \partial_t u_{VM}(t) \, dt = -g \cdot \int_0^\infty \partial_t u_{VM}(t) \, dt.$$

Hence, with the well-defined limits of u_{VM} at $t = 0$ and $t \to \infty$, we have

$$g \cdot \int_0^\infty \phi_{VM}(t) \, dt = -g \cdot \int_0^\infty \partial_t u_{VM}(t) \, dt = g \cdot (f - P_0(f)),$$

which yields the assertion since g is arbitrary.

In analogy to the classical linear setting, we would like to define a filtering of the wavelength representation via

$$\hat{u}_{filtered} = w_0 \, P_0(f) + \int_0^\infty w(t) \, \phi_{VM}(t) \, dt \tag{5.20}$$

for $w_0 \in \mathbb{R}$ and a suitable filter function $w(t)$. While the above formulation is the most intuitive expression for the filtering procedure, we have to take care of the regularity of ϕ_{VM} and hence understand the integral on the right-hand side in the sense of (5.18), i.e.,

$$\hat{u}_{filtered} = w_0 \, P_0(f) - \int_0^\infty (t w'(t) + w(t)) \, \partial_t u_{VM}(t) \, dt. \tag{5.21}$$

5.4.2 Scale Space Representation

A spectral representation based on the gradient flow formulation (5.1) was the first
work toward defining a nonlinear spectral decomposition and has been conducted
by the author of the book in [1, 2] for the case of J being the TV regularization. In
[3], the work was extended with colleagues to general absolutely one-homogeneous
functionals by observing that the solution of the gradient flow coincides with one of
the variational methods in the case of f being an eigenfunction, i.e., for $\|f\| = 1$,
$\lambda f \in \partial J(f)$, the solution to (5.1) is given by

$$u_{GF}(t) = \begin{cases} (1 - t\lambda)f & \text{for } t \leq \frac{1}{\lambda}, \\ 0 & \text{else.} \end{cases}$$

The latter motivates exactly the same definitions as for the variational method. In
particular, we define the wavelength decomposition of the input data f by

$$\phi_{GF}(t) = t\partial_{tt}u_{GF}(t).$$

As in the previous section, we will show that ϕ_{GF} also is a weak spectral wavelength decomposition in the sense of Definition 5.1 for any convex absolutely one-
homogeneous regularization functional J.

Proposition 5.3 (Finite time extinction) *Let J be an absolutely one-homogeneous
functional, and f be arbitrary. Then, there exists a time $T < \infty$ such that u_{GF}
determined via (5.1) meets*

$$u_{GF}(T) = P_0(f).$$

Proof Since any subgradient is orthogonal to $\mathcal{N}(J)$ the same holds for $\partial_t u$, hence
$P_0(u(t)) = P_0(u(0)) = P_0(f)$ for all $t \geq 0$. On the other hand, we see that

$$\frac{1}{2}\frac{d}{dt}\|Q_0 u_{GF}\|^2 = \langle Q_0 u_{GF}, Q_0 \partial_t u_{GF}\rangle = -\langle Q_0 u_{GF}, p_{GF}\rangle = -\langle u_{GF}, p_{GF}\rangle$$
$$= -J(u_{GF}) = -J(Q_0 u_{GF}) \leq -c_0\|Q_0 u_{GF}\|.$$

For t such that $\|Q_0 u_{GF}\| \neq 0$, we conclude

$$\frac{d}{dt}\|Q_0 u_{GF}\| \leq -c_0,$$

thus

$$\|Q_0 u_{GF}(t)\| \leq \|f\| - c_0 t.$$

Due to the positivity of the norm, we conclude that $\|Q_0 u_{GF}(T)\| = 0$ for $T \geq \frac{\|f\|}{c_0}$.

The regularity $\partial_t u_{GF}(t) \in L^\infty$ is guaranteed by the general theory on gradient
flows (cf. [8, p. 566, Theorem 3]); in our case, it can also be inferred quantitatively

from the a priori bound on the subgradients in Conclusion 1, Chap. 1. With the same proof as in Proposition 5.2, we can analyze ϕ_{GF} as a bounded linear functional:

Proposition 5.4 *The function* $u_{GF} : \mathbb{R}^+ \to \mathbb{R}^n$ *is Lipschitz continuous. Moreover, the spectral representation satisfies* $\phi_{GF} \in (W^{1,1}_{loc}(\mathbb{R}^+, \mathbb{R}^n))^*$.

Thus, we have the same regularity of the distribution as in the case of the variational method and can define filters in the same way.

Naturally, we obtain exactly the same reconstruction result as for the variational method.

Theorem 5.2 (Reconstruction of the input data) *It holds that*

$$f = P_0(f) + \int_0^\infty \phi_{GF}(t) \, dt. \tag{5.22}$$

Furthermore, due to the finite time extinction and the same smoothness of u_{GF} as for u_{VM}, we can define the formal filtering by

$$\hat{u}_{\text{filtered}} = w_0 P_0(f) - \int_0^\infty (tw'(t) + w(t)) \, \partial_t u_{GF}(t) \, dt, \tag{5.23}$$

for all filter functions $w \in W^{1,1}_{loc}$.

5.4.3 Inverse Scale Space Representation

A third way of defining a spectral representation proposed in [3] is via the inverse scale space flow Equation (2.4). Again, the motivation for the proposed spectral representation is based on the method's behavior on eigenfunctions. For $\|f\| = 1$, $\lambda f \in \partial J(f)$, the solution to (2.4) is given by

$$v_{IS}(s) = \begin{cases} 0 & \text{for } s \le \lambda, \\ f & \text{else.} \end{cases}$$

As we can see, the behavior of (2.4) is fundamentally different to the one of (5.16) and (5.1) in two aspects: First, stationarity is reached in an inverse fashion, i.e., by starting with zero and converging to f. Second, the primal variable $v_{IS}(s)$ has a piecewise constant behavior in time opposed to the piecewise linear behavior in the previous two cases. Naturally, only one derivative of v is necessary to obtain peaks. We define

$$\tilde{\phi}_{IS}(s) = \partial_s v_{IS}(s) = -\partial_{ss} q_{IS}(s)$$

to be the *frequency representation* of f in the inverse scale space setting.

We recall that we can relate frequency and wavelength representations by a change of variable $s = 1/t$ yielding

$$u_{IS}(t) = v_{IS}\left(\frac{1}{t}\right), \quad p_{IS}(t) = q_{IS}\left(\frac{1}{t}\right), \quad \phi_{IS}(t) = -\partial_s v_{IS}\left(\frac{1}{t}\right) = t^2 \partial_t u_{IS}(t).$$

(5.24)

Note that with these conversions we have

$$\int_0^\infty \phi(t) \cdot v(t) \, dt = \int_0^\infty \tilde\phi(s) \cdot v(1/s) \, ds,$$

and hence we may equally well consider integrations in the original variable s.

In the following, we will show that the above ϕ_{IS} is a weak spectral wavelength decomposition in the sense of Definition 5.1 for any convex absolutely one-homogeneous regularization functional J.

Analogous to the other methods, the inverse scale space methods have finite time extinction, and as we see from the proof even at the same time as the variational method:

Proposition 5.5 (Finite time extinction) *Let J be an absolutely one-homogeneous functional, and f be arbitrary. Then, there exists a time $T < \infty$ such that u_{IS} determined via (5.16) meets $u_{IS}(T) = P_0(f)$.*

Proof It is straightforward to see that $v_{IS}(s) = P_0(f)$ for $s \leq s_0$, where s_0 is the maximal value such that $s_0(f - P_0(f)) \in \partial J(0)$. The conversion to u_{IS} yields the finite time extinction.

Furthermore, note that the inverse scale space flow (2.4) can be written as a gradient flow on the dual variable $q_{IS}(s)$ with respect to the convex functional $J^*(q) - \langle f, q \rangle$. This guarantees that $\partial_s q_{IS}(s) \in L^\infty$, thus $v_{IS}(s) \in L^\infty$, and hence also $u_{IS}(t) \in L^\infty$.

Proposition 5.6 *The function $u_{IS} : \mathbb{R}^+ \to \mathbb{R}^n$ is bounded almost everywhere and $p_{IS} : [t_0, \infty) \to \mathbb{R}^n$ is Lipschitz continuous for every $t_0 > 0$. Moreover, the weak spectral representation satisfies $\phi_{IS} \in (W^{1,1}_{loc}(\mathbb{R}^+, \mathbb{R}^n))^*$.*

Proof The standard energy dissipation in the inverse scale space flow yields that $s \mapsto \|v_{IS}(s) - f\|$ is a nonincreasing function, and hence it is bounded by its value $\|f\|$ at $s = 0$. Hence, by the triangle inequality

$$\|v_{IS}(s)\| \leq 2\|f\|, \quad \|u_{IS}(t)\| \leq 2\|f\|$$

for all $s, t > 0$. The first inequality also implies the Lipschitz continuity of q_{IS} on $\mathbb{R}^+$ and hence by concatenation with $t \mapsto \frac{1}{t}$ Lipschitz continuity of p_{IS} on $[t_0, \infty)$.

We can now consider filterings of the inverse scale space flow representation, again by formal integration by parts

$$\hat{u}_{filtered} = \int_0^\infty w(t)\, \phi_{IS}(t)\, dt = -\int_0^\infty w\left(\frac{1}{s}\right) \partial_s v_{IS}(s)\, ds$$

$$= -\int_{\frac{1}{T}}^\infty w\left(\frac{1}{s}\right) \partial_s v_{IS}(s)\, ds = w_0 f - w_T P_0(f) - \int_0^T \frac{1}{s^2} w'\left(\frac{1}{s}\right) v_{IS}(s)\, ds$$

$$= w_0 f - w_T P_0(f) - \int_0^T w'(t)\, u_{IS}(t)\, dt,$$

where T is the finite extinction time stated by Proposition 5.5. The last line can be used to define filterings in the inverse scale space flow setting. Note that $w(t) = 1$ for all t leads to a reconstruction of the $Q_0 f$, i.e.,

$$f = P_0(f) + \int_0^\infty \phi_{IS}(t)\, dt. \tag{5.25}$$

5.4.4 Definitions of the Power Spectrum

As in the linear case, it is very useful to measure in some sense the "activity" at each frequency (scale). This can help identify dominant scales and design better the filtering strategies (either manually or automatically). Moreover, one can obtain a notion of the type of energy which is preserved in the new representation using an analog of Parseval's identity. While the amount of information on various spatial scales in linear and nonlinear scale spaces has been analyzed using Renyi's generalized entropies in [9], we will focus on defining a spectral power spectrum. As we have seen above, at least for an orthogonal spectral definition there is a natural definition of the power spectrum as the measure

$$S^2(t) = \Phi_t \cdot f =: S_3^2(t), \tag{5.26}$$

which yields a Parseval identity. Note that for S^2 being absolutely continuous with respect to the Lebesgue measure with density ρ, we can define a continuous power spectrum $s(t) = \sqrt{\rho(t)}$ and have

$$\|f\|^2 = \int_0^\infty s(t)^2\, dt. \tag{5.27}$$

On the other hand, if S^2 is a sum of concentrated measures, $S^2(t) = \sum_j a_j \delta(t - t_j)$ we can define $s_j = \sqrt{a_j}$ and have the classical Parseval identity

$$\|f\|^2 = \sum_j s_j^2. \tag{5.28}$$

We will now briefly discuss the three different spectrum definitions of (5.10), (5.11), and (5.13). For the sake of simplicity, we omit the subscripts VM, GF, and IS in the following discussion when all three variants can be used.

In [1, 2], a L^1 type spectrum was suggested for the TV spectral framework (without trying to relate to a Parseval identity),

$$S_1(t) := \|\phi(t)\|_1.$$

Considering the mathematical definition of ϕ as having components in $(W^{1,1}_{loc})^*$, we can see that mollification with a $W^{1,1}_{loc}$ function is needed in order to obtain a well-defined version of Eq. (5.10). A simple choice would be

$$S_1^\sigma(t) := \left\| \int_0^T g_\sigma(t)\, \phi(t; x)\, dt \right\|_{L^1(\Omega)} \tag{5.29}$$

for a Gaussian function $g_\sigma(t)$ with very small σ.

In [3], the following definition was suggested for the gradient flow,

$$S_2^2(t) := t^2 \frac{d^2}{dt^2} J(u_{GF}(t)) = 2t \langle \phi_{GF}(t), p_{GF}(t) \rangle = -t^2 \frac{d}{dt} \|p_{GF}(t)\|^2 \tag{5.30}$$

From the dissipation properties of the flow, one can deduce that S_2^2 is always non-negative. With this definition, the following analog of the Parseval identity can be shown:

$$\|f\|^2 = -\int_0^\infty \frac{d}{dt} \|u_{GF}(t)\|^2\, dt = 2\int_0^\infty \langle p_{GF}(t), u_{GF}(t) \rangle\, dt = 2\int_0^\infty J(u_{GF}(t))\, dt$$

$$= -2\int_0^\infty t\frac{d}{dt} J(u_{GF}(t))\, dt = \int_0^\infty t\frac{d}{dt} J(u_{GF}(t))\, dt = \int_0^\infty S_2(t)^2\, dt. \tag{5.31}$$

Below, we will show that under certain conditions indeed S_3^2 is an equivalent realization of S_2^2.

5.5 Analysis of the Spectral Decompositions

5.5.1 Basic Conditions on the Regularization

To analyze the behavior and relation of the different spectral decompositions, let us make the following definition. We note that this analysis is in finite dimensions $(u \in R^n)$.

Definition 5.3 (*MINSUB*) We say that J meets (MINSUB) if for all $u \in R^n$, the element $\hat{p}$ determined by

$$\hat{p} = \arg\min_p \|p\|^2 \text{ subject to } p \in \partial J(u), \tag{5.32}$$

meets

$$\langle \hat{p}, \hat{p} - q \rangle = 0 \ \forall q \in \partial J(u).$$

To give some intuition about what the condition (MINSUB) means, let us give two examples of regularizations that meet (MINSUB).

Example 5.1 (ℓ^1 *regularization meets* (*MINSUB*)) Consider $J(u) = \|u\|_1$. The characterization of the ℓ^1 subdifferential yields

$$q \in \partial J(u) \quad \Leftrightarrow \quad q_l \begin{cases} = 1 & \text{if } u_l > 0, \\ = -1 & \text{if } u_l < 0, \\ \in [-1, 1] & \text{if } u_l = 0. \end{cases} \tag{5.33}$$

Consequently, the $\hat{p}$ defined by Eq. (5.32) meets $\hat{p}_l = 0$ for all l with $u_l = 0$. Consider any other $q \in \partial J(u)$. Then,

$$\langle \hat{p}, \hat{p} - q \rangle = \underbrace{\sum_{l, u_l > 0} \underbrace{\hat{p}_l}_{=1} (\underbrace{\hat{p}_l}_{=1} - q_l)}_{=0} + \underbrace{\sum_{l, u_l < 0} \underbrace{\hat{p}_l}_{=-1} (\underbrace{\hat{p}_l}_{=-1} - q_l)}_{=0} + \underbrace{\sum_{l, u_l = 0} \underbrace{\hat{p}_l}_{=0} (\hat{p}_l - q_l)}_{=0} = 0,$$

which shows that $J(u) = \|u\|_1$ meets (MINSUB).

Another interpretation of the (MINSUB) condition is geometric. Since in Hilbert spaces the scalar product being zero expresses orthogonality, one could also think of (MINSUB) requiring the subdifferentials of J being well-behaved polyhedrons (or being single valued). If the faces of polyhedron representing all possible subgradients are oriented such that they are tangent to the circle with radius $\|\hat{p}\|$, then (MINSUB) is met. An interesting possible connection to investigate in future research is the connection of (MINSUB) to the notion of *minimal norm certificates* in the field of compressive sensing [10]. Recent studies use the certificate to prove robustness to noise of the regularizer and model consistency [11, 12].

A second property we are interested in is a particular behavior of the flow that allows to state that the spectral representation merely consists of a collection of δ-peaks. For this, we need the definition of a polyhedral seminorm.

Definition 5.4 (*Polyhedral Seminorm* (*PS*)) We say that J induces a *polyhedral seminorm* (PS) if there exists a finite set whose convex hull equals $\partial J(0)$.

Polyhedral seminorms have several interesting properties:

Proposition 5.7 *Let J satisfy (PS), then the set*

$$\{\arg\min_p \|p\|^2, \quad such\ that\ p \in \partial J(u) \mid u \in \mathscr{X}\}$$

is finite.

Proof By Lemma 1.3, we know that $\partial J(u)$ is the intersection of the polyhedral shape $\partial J(0)$ and the linear manifold defined by $\langle p, u \rangle = J(u)$. Since the intersection cannot contain any interior part of $\partial J(0)$ and due to (PS), the intersection must be a face of $\partial J(0)$. Hence, the set of minimizers of $\|p\|$ in the subgradients of $\partial J(u)$ is contained in the set of minimizers of $\|p\|$ on single face of $\partial J(0)$. Since the number of faces is finite by (PS), also the set of minimizers (for each face, there is a unique one) is finite.

As an example, consider the ℓ^1 norm on $\mathbb{R}^n$ again. The set $\partial \|0\|_1$ coincides with the unit ℓ^∞ ball and therefore is polyhedral. The set considered in Proposition 5.7 is the collection of all p such that the ith component of p is in $\{-1, 1, 0\}$, which is of course finite. Note that all regularizations of the form $J(u) = \|Ku\|_1$, for an arbitrary linear operator K, are also polyhedral as $\partial J(0) = K^T \partial \|0\|_1$. Interestingly, the most general form of a J meeting (PS) is in fact $J(u) = \|Ku\|_\infty$:

Proposition 5.8 *Let an absolutely one-homogeneous function $J : \mathbb{R}^n \to \mathbb{R}$ meet (PS). Then, there is a matrix P such that $J(u) = \|Pu\|_\infty$.*

Proof Let $u \in \mathbb{R}^n$ be arbitrary, and let $p \in \partial J(u)$. By (PS), we can write p as a convex combination of finitely many subgradients $p_i \in \partial J(0)$, $p = \sum_i \alpha_i p_i$. We find

$$J(u) = \langle u, p \rangle = \sum_i \alpha_i \langle u, p_i \rangle \leq \sum_i \alpha_i J(u) = J(u).$$

The fact that equality holds in the above estimate shows that $p_i \in \partial J(u)$ for all i with $\alpha_i > 0$, such that $J(u) = \max_i \langle p_i, u \rangle$. Since $J(u) = J(-u)$ by the absolute one-homogenity, we may as well consider $J(u) = \max_i |\langle p_i, u \rangle|$, or, after writing the p_i as rows of a matrix P, $J(u) = \|Pu\|_\infty$.

To be able to state our main results compactly, let us introduce a third abbreviation in addition to (PS) and (MINSUB).

Definition 5.5 *(DDL1)* We refer to the specific assumption that we are considering diagonally dominant ℓ^1 regularization, i.e., the particular case where $J(u) = \|Ku\|_1$ for a matrix K such that KK^* is diagonally dominant, as (DDL1).

To provide some more intuition for (DDL1), please note that the discrete total variation regularization in one dimension meets (DDL1), since the corresponding matrix KK^* has the entry 2 on the diagonal and at most two entries equal to -1 on the off-diagonal. The discrete anisotropic total variation in two dimensions, however, does not meet (DDL1) as the largest value on the diagonal of KK^* is still 2 but there are up to 6 off-diagonal entries equal to 1 or -1.

5.5.2 Connection Between Spectral Decompositions

In this section, we will analyze the relation between the different approaches (5.16), (5.1), and (2.4) and their induced spectral decompositions for polyhedral regularizations and the smoothness assumptions of (MINSUB) or (DDL1). Note that several of the results stated below have been observed in particular cases, such as the equivalence of the variational approach and the scale space flow for 1D total variation regularization [13], the piecewise linear behavior of ℓ^1 regularization used in the celebrated homotopy algorithm [14], and several extensions, for instance, for analysis sparsity [15] and ℓ^∞ regularization in [10].

Using the three assumptions and definitions (PS), (MINSUB), and (DDL1) introduced in the previous subsection, we can state the following results.

Theorem 5.3 (Piecewise dynamics for J satisfying (PS)) *If J meets (PS), then the scale space flow, the inverse scale space flow, and the variational method have piecewise linear, respectively, piecewise constant dynamics. In more detail:*

- *There exist finitely many $0 = t_0 < t_1 < t_2 < \cdots < t_K = \infty$ such that the solution of (5.1) is given by*

$$u_{GF}(t) = u_{GF}(t_i) - (t - t_i)p_{GF}(t_{i+1}), \qquad (5.34)$$

  for $t \in [t_i, t_{i+1}]$. In other words $p_{GF}(t_{i+1}) \in \partial J(u_{GF}(t))$ for $t \in [t_i, t_{i+1}]$ such that the dynamics of the flow is piecewise linear in $u_{GF}(t)$ and piecewise constant in $p_{GF}(t)$.
- *There exist finitely many $0 < s_1 < s_2 < \cdots < s_L = \infty$ such that the solution of (2.4) is given by*

$$q_{IS}(s) = q_{IS}(s_i) - (s - s_i)(f - v_{IS}(s_{i+1})), \qquad (5.35)$$

  for $s \in [s_i, s_{i+1}]$. In other words $q_{IS}(s) \in \partial J(v_{IS}(s_{i+1}))$ for $s \in [s_i, s_{i+1}]$ such that the dynamics of the flow is piecewise linear in $q_{IS}(t)$ and piecewise constant in $v_{IS}(t)$.
- *There exist finitely many $0 < t_1 < t_2 < \cdots < t_L = \infty$ such that the solution of (5.16) satisfies that u_{VM} is an affinely linear function of t for $t \in [t_i, t_{i+1}]$, and for $p_{VM}(t) \in \partial J(u_{VM}(t))$ denoting the corresponding subgradient, $q_{VM}(s) = p_{VM}(1/s)$ is an affinely linear function of s for $s \in [s_i, s_{i+1}]$, $s_i = \frac{1}{t_i}$.*

Proof (Gradient Flow) Let us begin with the proof of the piecewise linear dynamics of the scale space flow. For any $t_i \geq 0$ (starting with $t_0 = 0$ and $u_{GF}(0) = f$), we consider

$$p_{GF}(t_{i+1}) = \arg \min_p \|p\|^2 \quad \text{such that } p \in \partial J(u_{GF}(t_i)) \qquad (5.36)$$

and claim that $p_{GF}(t_{i+1}) \in \partial J(u_{GF}(t_i) - (t - t_i)p_{GF}(t_{i+1}))$ for small enough $t > t_i$. Due the property of J having finitely representable subdifferentials

(Proposition 5.7), we know that for any t there has to exist a q^j in the finite set M such that $q^j \in \partial J(u_{GF}(t_i) - (t - t_i)p_{GF}(t_{i+1}))$. The characterization of the subdifferential of absolutely one-homogeneous functions (1.20) tells us that $q^j \in \partial J(u_{GF}(t_i) - (t - t_i)p_{GF}(t_{i+1}))$ is met by those q^j which maximize $\langle q^j, u_{GF}(t_i) - (t - t_i)p_{GF}(t_{i+1}) \rangle$. We distinguish two cases:

- First of all consider $q^j \in \partial J(u_{GF}(t_i))$. In this case, the optimality condition to (5.36) tells us that

$$\langle p_{GF}(t_{i+1}), p_{GF}(t_{i+1}) - q \rangle \leq 0 \quad \forall q \in \partial J(u_{GF}(t_i)), \tag{5.37}$$

and we find

$$\begin{aligned}
\langle u_{GF}(t_i) - (t - t_i)p_{GF}(t_{i+1}), q^j \rangle &= J(u_{GF}(t_i)) - (t - t_i)\langle p_{GF}(t_{i+1}), q^j \rangle, \\
&= \langle u_{GF}(t_i), p_{GF}(t_{i+1}) \rangle - (t - t_i)\langle p_{GF}(t_{i+1}), q^j \rangle, \\
&= \langle u_{GF}(t_i) - (t - t_i)p_{GF}(t_{i+1}), p_{GF}(t_{i+1}) \rangle \\
&\quad + (t - t_i)\langle p_{GF}(t_{i+1}), p_{GF}(t_{i+1}) - q^j \rangle, \\
&\overset{(5.37)}{\leq} \langle u_{GF}(t_i) - (t - t_i)p_{GF}(t_{i+1}), p_{GF}(t_{i+1}) \rangle.
\end{aligned}$$

- For $q^j \notin \partial J(u_{GF}(t_i))$ we can use Remark 1.1, define $c^j(t_i) := J(u_{GF}(t_i)) - \langle q^j, u_{GF}(t_i) \rangle > 0$. We compute

$$\begin{aligned}
\langle u_{GF}(t_i) - (t - t_i)p_{GF}(t_{i+1}), q^j \rangle &= \langle u_{GF}(t_i), q^j \rangle - (t - t_i)\langle p_{GF}(t_{i+1}), q^j \rangle \\
&= J(u_{GF}(t_i)) - c^j(t_i) - (t - t_i)\langle p_{GF}(t_{i+1}), q^j \rangle \\
&= J(u_{GF}(t_i)) - (t - t_i)\|p_{GF}(t_{i+1})\|^2 \\
&\quad - \underbrace{\left(c^j(t_i) - (t - t_i)(\|p_{GF}(t_{i+1})\|^2 - \langle p_{GF}(t_{i+1}), q^j \rangle) \right)}_{\geq 0 \text{ for small } t > t_i} \\
&\leq \langle u_{GF}(t_i) - (t - t_i)p_{GF}(t_{i+1}), p_{GF}(t_{i+1}) \rangle \quad \text{for small } t.
\end{aligned}$$

Since the set of all q^j is finite, we can conclude that there exists a smallest time $t > t_i$ over all j up until which $p_{GF}(t_{i+1}) \in \partial J(u_{GF}(t_i) - (t - t_i)p_{GF}(t_{i+1}))$.

To see that the number of times t_k at which the flow changes is finite, recall that each $p_{GF}(t_{i+1})$ is determined via Eq. (5.36) and that the corresponding minimizer is unique. Since $p_{GF}(t_i)$ is in the feasible set, we find $\|p_{GF}(t_{i+1})\| < \|p_{GF}(t_i)\|$ (unless $p_{GF}(t_i) = p_{GF}(t_{i+1})$ in which case the time step t_i was superfluous to consider). Since the number of possible $p_{GF}(t_i)$ is finite by Proposition 5.7, and the property $\|p_{GF}(t_{i+1})\| < \|p_{GF}(t_i)\|$ shows that each $p_{GF}(t_i)$ be attained at most once, the number of t_i at which the flow changes is finite. This concludes the proof for the scale space case.

Inverse Scale Space Flow. Similar to the above proof, for any $s_i \geq 0$ (starting with $s_0 = 0$ and $q_{IS}(0) = 0$), we consider

$$v_{IS}(s_{i+1}) = \arg\min_{v} \|v - f\|^2 \quad \text{such that } v \in \partial J^*(q_{IS}(s_i))$$

and claim that $q_{IS}(s_i) + (s - s_i)(f - v_{IS}(s_{i+1})) \in \partial J(v_{IS}(s_{i+1}))$ for small enough $s > s_i$.

The optimality condition to the above minimization problem is

$$\langle f - v_{IS}(s_{i+1}), v_{IS}(s_{i+1}) - v \rangle \geq 0 \quad \forall v \in \partial J^*(q_{IS}(s_i)). \tag{5.38}$$

By choosing $v = 0$ and $v = 2v_{IS}(s_{i+1})$, we readily find

$$\langle f - v_{IS}(s_{i+1}), v_{IS}(s_{i+1}) \rangle = 0,$$

and therefore

$$\langle q_{IS}(s_i) + (s - s_i)(f - v_{IS}(s_{i+1})), v_{IS}(s_{i+1}) \rangle = J(v_{IS}(s_{i+1}))$$

for all t. Furthermore, (5.38) reduces to

$$\langle f - v_{IS}(s_{i+1}), v \rangle \leq 0 \quad \forall v \in \partial J^*(q_{IS}(s_i)). \tag{5.39}$$

It is hence sufficient to verify that

$$q_{IS}(s_i) + (s - s_i)(f - v_{IS}(s_{i+1})) \in \partial J(0)$$

for $s > s_i$ sufficiently small. It is well known (cf. [16]) that any set that is represented as the convex hull of finitely many points can also be represented as the intersection of finitely many half-spaces, such that

$$\partial J^*(0) = \{q \mid Bq \leq c\}$$

for some matrix B and vector c. Let b_l denote the rows of the matrix B. Since $\langle b_l, q \rangle = J(b_l)$ for $q \in \partial J(b_l)$, it is obvious that $J(b_l) \leq c_l$. We distinguish two cases:

- If $b_l \in \partial J^*(q_{IS}(s_i))$, then (5.39) yields

$$\langle q_{IS}(s_{i+1}) + (s - s_i)(f - v_{IS}(s_{i+1})), b_l \rangle = J(b_l) + (s - s_i)\langle f - v_{IS}(s_{i+1}), b_l \rangle \leq J(b_l)$$

- If $b_l \notin \partial J^*(q_{IS}(s_i))$, then $\langle q_{IS}(s_i), b_l \rangle < J(b_l)$, such that

$$\langle q_{IS}(s_i) + (s - s_i)(f - v_{IS}(s_{i+1})), b_l \rangle \leq J(b_l)$$

for $s \in [s_i, s_{i+1}^l]$ with $s_{i+1}^l > s_i$ small enough.

We define $s_{i+1} = \min_l s^l_{i+1}$, which yields the piecewise constant behavior. Similar to the proof for the gradient flow, the $v_{IS}(s_{i+1})$ are unique and $v_{IS}(s_i)$ is in the feasible set, such that $\|f - v_{IS}(s_{i+1})\| < \|f - v_{IS}(s_i)\|$. Since the total number of possible constraints $v \in \partial J^*(p)$ is finite, there can only be finitely many times s_i at which the flow changes.

Variational Method. Since q_{VM} is a Lipschitz continuous function of s, the subgradient changes with finite speed on $\partial J(0)$. This means that p_{VM} needs finite time to change from a facet to another facet of the polyhedral set $\partial J(0)$; in other words, there exist times $0 < \tilde{s}_1 < \tilde{s}_2 < \cdots < \tilde{s}_L = \infty$ such that $q_{VM}(\tilde{s}_i) \in \partial J(u_{VM}(s))$ for $s \in [\tilde{s}_i, \tilde{s}_{i+1})$. Comparing the optimality condition for a minimizer at time s and time $\tilde{s}_i$, we find

$$s\left(v_{VM}(s) - \frac{\tilde{s}_i}{s}v_{VM}(s_i) - \frac{s - \tilde{s}_i}{s}f\right) + q_{VM}(s) - q_{VM}(\tilde{s}_i) = 0.$$

Now let v be any element such that $q_{VM}(\tilde{s}_i) \in \partial J(v)$. Then, we have

$$\langle v_{VM}(s) - \frac{\tilde{s}_i}{s}v_{VM}(\tilde{s}_i) - \frac{s - \tilde{s}_i}{s}f, v - v_{VM}(s)\rangle = \frac{1}{s}\langle q_{VM}(s) - q_{VM}(\tilde{s}_i), v_{VM}(s) - v\rangle \geq 0.$$

This implies that $v_{VM}(s)$ is the minimizer

$$v_{VM}(s) = \arg\min_v \left\| v - \frac{\tilde{s}_i}{s}v_{VM}(\tilde{s}_i) - \frac{s - \tilde{s}_i}{s}f \right\|^2 \qquad \text{such that } q_{VM}(\tilde{s}_i) \in \partial J(v),$$

i.e., it equals the projection of $\frac{\tilde{s}_i}{s}v_{VM}(\tilde{s}_i) + \frac{s-\tilde{s}_i}{s}f$ on the set of v such that $q_{VM}(\tilde{s}_i) \in \partial J(v)$, which is an intersection of a finite number of half-spaces. Since $u_{VM}(\frac{1}{s})$ is an affinely linear function for $s \in [\tilde{s}_i, \tilde{s}_{i+1}]$, its projection to the intersection of a finite number of half-spaces is piecewise affinely linear for $s \in [\tilde{s}_i, \tilde{s}_{i+1}]$. Hence, the overall dynamics of u_{VM} is piecewise affinely linear in t. The piecewise affine linearity of p_{VM} in terms of $s = \frac{1}{t}$ then follows from the optimality condition by a direct computation.

Note that the results regarding the scale space and inverse scale space flows were to be expected based on the work [17] on polyhedral functions. The above notation and the proof is, however, much more accessible since it avoids the lengthy notation of polyhedral and finitely generated functions.

From Theorem 5.3, we can draw the following simple but important conclusion.

Conclusion 3 *If J meets (PS), then the scale space flow, the inverse scale space flow, and the variational method have a well-defined spectral representation, which consists of finitely many δ-peaks. In other words, the spectral representation is given by*

$$\phi_*(t) = \sum_{i=0}^{N_*} \phi_*^i \delta(t - t_i), \qquad for * \in \{VM, GF, IS\}, \qquad (5.40)$$

and the reconstruction formulas (5.19), (5.22), or (5.25), yield a decomposition of f as

$$f = \sum_{i=0}^{N_*} \phi_*^i, \qquad for * \in \{VM, GF, IS\}, \qquad (5.41)$$

where t_i are the finite number of times where the piecewise behavior of the variational method, the gradient flow, or the inverse scale space flow stated in Theorem 5.3 changes. The corresponding ϕ_*^i can be seen as multiples of $\psi_*(t)$ arising from the polar decomposition of the spectral frequency representation. They are given by

$$\phi_{GF}^i = t_i(p_{GF}(t_i) - p_{GF}(t_{i+1})),$$
$$\phi_{VM}^i = t_i(u_{VM}(t_{i+1}) - 2u_{VM}(t_i) + u_{VM}(t_{i-1})),$$
$$\phi_{IS}^i = u_{IS}(t_i) - u_{IS}(t_{i+1}),$$

with $u_{IS}(t_0) = u_{VM}(t_0) = u_{VM}(t_{-1}) = f$, $p_{GF}(t_0) = 0$, $t_0 = 0$.

Naturally, we should ask what the relation between the different spectral decomposition methods proposed earlier. We can state the following results:

Theorem 5.4 (Equivalence of GF and VM under (MINSUB)) *Let J be such that (PS) and (MINSUB) are satisfied. Then,*

1. *$p_{GF}(s) \in \partial J(u_{GF}(t))$ for all $t \geq s$.*
2. *The solution $u_{GF}(t)$ meets $u_{GF}(t) = u_{VM}(t)$ for all t where $u_{VM}(t)$ is a solution of (5.16). The relation of the corresponding subgradients is given by*

$$p_{VM}(t) = \frac{1}{t}\int_0^t p_{GF}(s)ds \in \partial J(u_*(t)), \qquad (5.42)$$

for $* \in \{VM, GF\}$.

Proof Based on Theorem 5.3, we know that there exist times $0 < t_1 < t_2 < \ldots$ in between which $u(t)$ behaves linearly. We proceed inductively. For $0 \leq t \leq t_1$, we have

$$u_{GF}(t) = u_{GF}(0) - t\, p_{GF}(t_1) = f - t\, p_{GF}(t_1),$$

with $p_{GF}(t_1) \in \partial J(u_{GF}(t))$ for all $t \in [0, t_1]$. The latter coincides with the optimality condition for $u_{GF}(t) = argmin_u \frac{1}{2}\|u - f\|_2^2 + tJ(u)$, which shows $u_{GF}(t) = u_{VM}(t)$ for $t \in [0, t_1]$. Due to the closedness of subdifferentials, $p_{GF}(t_1) \in \partial J(u(t))$ for $t \in [0, t_1[$, implies that the latter holds for $t = t_1$ too. Thus, we can state that $p_{GF}(t_1) = p_{VM}(t) = \frac{1}{t}\int_0^t p_{GF}(t)\,dt \in \partial J(u_*(t))$ for $t \in [0, t_1]$, $* \in \{GF, VM\}$.

Assume the assertion holds for all $t \in [0, t_i]$. We will show that it holds for $t \in [0, t_{i+1}]$, too. Based on the proof of Theorem 5.3, we know that $u_{GF}(t) = u_{GF}(t_i) + (t - t_i)p_{GF}(t_{i+1})$ for $t \in [t_i, t_{i+1}]$, and $p_{GF}(t_{i+1}) = argmin_{p \in \partial J(u(t_i))} \|p\|_2^2$. Now (MINSUB) implies $\langle p_{GF}(t_{i+1}), p_{GF}(t_{i+1}) - p_{GF}(t_j) \rangle = 0$ for all $j \le i$. We compute

$$
\begin{aligned}
\langle p_{GF}(t_j), u_{GF}(t) \rangle &= \langle p_{GF}(t_j), u_{GF}(t_i) + (t - t_i)p_{GF}(t_{i+1}) \rangle \\
&= J(u_{GF}(t_i)) + (t - t_i)\langle p_{GF}(t_j), p_{GF}(t_{i+1}) \rangle \\
&= \langle p_{GF}(t_{i+1}), u(t_i) \rangle + (t - t_i)\langle p_{GF}(t_{i+1}), p_{GF}(t_{i+1}) \rangle \\
&= \langle p_{GF}(t_{i+1}), u(t_i) + (t - t_i)p_{GF}(t_{i+1}) \rangle \\
&= \langle p_{GF}(t_{i+1}), u(t) \rangle \\
&= J(u(t)).
\end{aligned}
$$

Therefore, $p_{GF}(t_j) \in \partial J(u(t))$ for all $t \in [t_i, t_{i+1}]$ and all $j \le i + 1$. Integrating the scale space flow equation yields for $t \le t_{i+1}$

$$
0 = u_{GF}(t) - f + \int_0^t p_{GF}(t) \, dt = u_{GF}(t) - f + t \left(\frac{1}{t} \int_0^t p_{GF}(t) \, dt \right).
$$

Due to the convexity of the subdifferential, we find $\frac{1}{t} \int_0^t p_{GF}(t) \, dt \in \partial J(u_{GF}(t))$, which shows that $u_{GF}(t)$ meets the optimality condition for the variational method and concludes the proof.

The above theorem not only allows us to conclude that $\phi_{VM} = \phi_{GF}$ but also $\phi_{VM}^i = \phi_{GF}^i$ for all i in the light of Eq. (5.41). Additionally, the first aspect of the above theorem allows another very interesting conclusion, revealing another striking similarity to linear spectral decompositions.

5.5.3   Orthogonality of the Spectral Components

We will now show that under certain conditions we can prove that the spectral components $\phi(t)$ are orthogonal to each other. This is a major result obtained so far in the analysis of the nonlinear spectral representation. It holds precisely in the discrete one-dimensional case for TV for instance. This is shown in Figs. 5.4 and 5.5 where a 1D experiment is conducted. The TV-flow solution $u(t)$ is shown along with the corresponding subgradient $p(t)$ and spectral component $\phi(t)$. We will see later that p are eigenfunctions in this case. However, the more interesting and useful entity is ϕ. This is different than the linear case, where the eigenfunctions composing the signal are orthogonal. Here, the *differences of two eigenfunctions*, which are the ϕ components, are orthogonal to each other. In Fig. 5.5, we see the correlation of the five elements which compose the input signal f. Whereas the p components are highly correlated, and therefore are not adequate to be used for filtering, the ϕ components

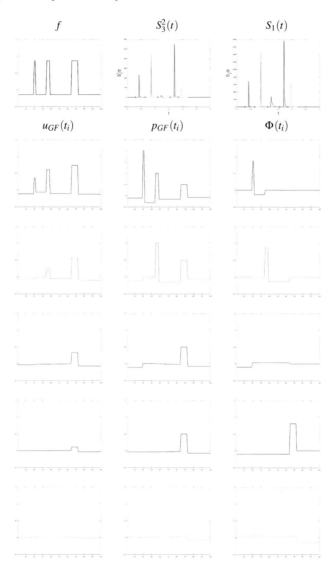

Fig. 5.4 Decomposition example. An illustration of Theorem 5.8. Top (from left): input signal f, spectrum $S_3^2(t)$ and $S_1(t)$. From second to sixth row, $u_{GF}(t_i)$ (left), $p_{GF}(t_i)$ (center), and $\Phi(t_i)$ at time points marked by circles in the spectrum plots. $\Phi(t_i)$ is an integration of $\phi_{GF}(t)$ between times t_i and t_{i+1}, visualized with different colors in the spectrum plots

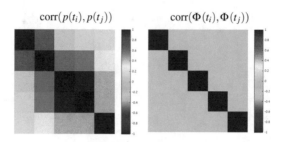

Fig. 5.5 Illustration of Theorem 5.5 (Orthogonal Decompositions). Correlation matrix of the p (left) and the Φ elements in the example shown in Fig. 5.4. Whereas the p elements are correlated, the Φ elements are very close numerically to being orthogonal

are uncorrelated. This essentially means that one can attenuating or amplifying any component separately (without creating major artifacts). In the 2D case, we see strong indications that the ϕ components are highly uncorrelated numerically, but currently we do not have the theoretical basis to support this.

Theorem 5.5 (Orthogonal Decompositions) *Let J be such that (PS) and (MINSUB) are satisfied. Then, Eq. (5.41) is an orthogonal decomposition of the input data f for the variational method as well as for the gradient flow, i.e.,*

$$\langle \phi_*^i, \phi_*^j \rangle = 0, \qquad \forall i \neq j, \ * \in \{VM, GF\}$$

Proof Due to the equivalence stated in Theorem 5.4, it is sufficient to consider the gradient flow only. For the gradient flow, we have $\phi_{GF}^i = t_i(p_{GF}(t_i) - p_{GF}(t_{i+1}))$. Let $i > j$, then

$$\langle \phi_*^i, \phi_*^j \rangle = t_i t_j \langle p_{GF}(t_{i+1}) - p_{GF}(t_i), p_{GF}(t_{j+1}) - p_{GF}(t_j) \rangle$$
$$= t_i t_j \left(\langle p_{GF}(t_{i+1}), p_{GF}(t_{j+1}) - p_{GF}(t_j) \rangle - \langle p_{GF}(t_i), p_{GF}(t_{j+1}) - p_{GF}(t_j) \rangle \right).$$

Now due to statement 1 in Theorem 5.4, we know that $p_{GF}(t_j) \in \partial J(u(t^i))$ and $p_{GF}(t_{j+1}) \in \partial J(u(t^i))$. Since $p_{GF}(t_i)$ was determined according to Eq. (5.36), we can use (MINSUB) to state that

$$\langle p_{GF}(t_i), p_{GF}(t_{j+1}) \rangle = \|p_{GF}(t_i)\|^2 = \langle p_{GF}(t_i), p_{GF}(t_j) \rangle,$$

which means that $\langle p_{GF}(t_i), p_{GF}(t_{j+1}) - p_{GF}(t_j) \rangle = 0$. With exactly the same argument, we find $\langle p_{GF}(t_{i+1}), p_{GF}(t_{j+1}) - p_{GF}(t_j) \rangle = 0$ which yields the assertion.

Due to the limited number of possibly orthogonal vectors in $\mathbb{R}^n$, we can conclude the following:

Conclusion 4 *For J : $\mathbb{R}^n \rightarrow \mathbb{R}$ meeting (PS) and (MINSUB), the number of possible times at which the piecewise behavior of the variational and scale space method changes is at most n.*

As announced before, we also obtain the equivalence of definitions of the power spectrum:

Proposition 5.9 *Let J be such that (PS) and (MINSUB) are satisfied. Then for the spectral representations of the gradient flow and the variational method, we have $S^2(t) = S_2^2(t)$ for almost every t (and S_2^2 being defined via (5.11)).*

Proof Due to the equivalence of representations, it suffices to prove the result for the gradient flow case. We have due to the orthogonality of the ϕ_{GF}^i

$$S^2(t) = \sum_{i=0}^{N_*}(\phi_{GF}^i \cdot f)\delta(t - t_i) = \sum_{i=0}^{N_*}\|\phi_{GF}^i\|^2\delta(t - t_i)$$

$$= \sum_{i=0}^{N_*}t_i^2(p_{GF}(t_i) - p_{GF}(t_{i+1})) \cdot (p_{GF}(t_i) - p_{GF}(t_{i+1}))\delta(t - t_i).$$

With (MINSUB), we conclude

$$p_{GF}(t_i) \cdot p_{GF}(t_{i+1}) = \|p_{GF}(t_{i+1})\|^2.$$

Inserting this relation, we have

$$S^2(t) = \sum_{i=0}^{N_*}t_i^2(\|p_{GF}(t_i)\|^2 - \|p_{GF}(t_{i+1})\|^2)\delta(t - t_i) = t^2\frac{d}{dt}\|p(t)\|^2 = S_2^2(t).$$

Due to the importance of (MINSUB) based on Theorems 5.4 and 5.5, the next natural question is if we can give a class of regularization functionals that meet (MINSUB).

Theorem 5.6 *Let (DDL1) be met (Definition 5.5). Then J meets (PS) and (MINSUB).*

Proof First of all, for any $p \in \partial J(u)$ we know from the characterization of the subdifferential (5.33) of the ℓ^1 norm that $p = K^T q$ with

$$q(l) \begin{cases} = 1 & \text{if } Ku(l) > 0, \\ = -1 & \text{if } Ku(l) < 0, \\ \in [-1, 1] & \text{if } Ku(l) = 0. \end{cases} \tag{5.43}$$

Thus for any given $\hat{u}$ and $\hat{p}$ defined as

$$\hat{p} = \arg\min_{p} \|p\|^2 \text{ s.t. } p \in \partial J(\hat{u}) \tag{5.44}$$

we have the optimality condition

$$KK^T\hat{q} + \lambda = 0, \tag{5.45}$$

for some Lagrange multiplier λ to enforce the constraints. For a given u denote

$$I_u = \{l \mid Ku(l) \neq 0\}. \tag{5.46}$$

For better readability of the proof, let us state an additional lemma:

Lemma 5.1 *Let $J(u) = \|Ku\|_1$ for a linear operator K and let $\hat{p}$ be defined by (5.44) for some arbitrary element u. If the λ arising from (5.45) meets $\lambda(l) = 0 \; \forall l \notin I_u$, then (MINSUB) holds.*

Proof Let $z \in \partial J(u)$ be some other subgradient. Based on (5.33), there is a q_z such that $z = K^T q_z$. Now

$$\langle \hat{p}, \hat{p} - z \rangle = \langle KK^T \hat{q}, \hat{q} - q_z \rangle = \sum_l (KK^T \hat{q})(l) \cdot (\hat{q}(l) - q_z(l))$$

$$= \sum_{l \in I} (KK^T \hat{q})(l) \cdot (\hat{q}(l) - q_z(l)) + \sum_{l \notin I} (KK^T \hat{q})(l) \cdot (\hat{q}(l) - q_z(l))$$

$$= \sum_{l \notin I} (KK^T \hat{q})(l) \cdot (\hat{q}(l) - q_z(l)) = -\sum_{l \notin I} \lambda(l) \cdot (\hat{q}(l) - q_z(l)) = 0$$

Consider $\hat{p}$ defined by (5.44) for some arbitrary element u. According to Lemma 5.1, it is sufficient to show that the Lagrange multiplier λ in (5.45) meets $\lambda(l) = 0 \; \forall l \notin I_u$. Assume $\lambda(l) > 0$ for some $l \notin I$. The complementary slackness condition then tells us that $\hat{q}(l) = 1$. Therefore,

$$-\lambda(l) = (K^T K \hat{q})(l), = \sum_j (KK^T)(l, j) \hat{q}(j),$$

$$= (KK^T)(l, l) \hat{q}(l) + \sum_{j \neq l} (KK^T)(l, j) \hat{q}(j), = (KK^T)(l, l) + \sum_{j \neq l} (KK^T)(l, j) \hat{q}(j)$$

$$\geq (KK^T)(l, l) - \sum_{j \neq l} |(KK^T)(l, j)| \stackrel{KK^T \text{ diag. dom.}}{\geq} 0,$$

which is a contradiction to $\lambda(l) > 0$. A similar computation shows that $\lambda(l) < 0$ (which implies $\hat{q}(l) = -1$) is not possible either.

By Theorem 5.4, the above result implies that all (DDL1) regularizations lead to the equivalence of the spectral representations obtained by the variational and the scale space method. Interestingly, the class of (DDL1) functionals also allows to show the equivalence of the third possible definition of a spectral representation.

Theorem 5.7 (Equivalence of *GF*, *VM*, and *IF*) *Let (DDL1) be met. Denote $\tau = \frac{1}{t}$, $v(\tau) = u_{GF}(1/\tau) = u_{GF}(t)$, and $r(\tau) = p_{VM}(1/\tau) = p_{VM}(t)$. It holds that*

$$\partial_\tau r(\tau) = f - \partial_\tau (\tau \, v(\tau)), \quad r(\tau) \in \partial J (\partial_\tau (\tau \, v(\tau))), \tag{5.47}$$

in other words, $(r(\tau), \partial_\tau (\tau \, v(\tau)))$ solve the inverse scale space flow (2.4).

Proof First of all note that

$$\tau(u_{VM}(\tau) - f) + p_{VM}(\tau) = 0$$

holds as well as $u_{VM}(\tau) = u_{GF}(\tau)$ based on Theorem 5.4. Differentiating the above equality yields

$$\partial_\tau r(\tau) = f - \partial_\tau(\tau v(\tau)).$$

We still need to show the subgradient inclusion. It holds due to the piecewise linearity of the flow that

$$\partial_\tau(\tau v(\tau)) = v(\tau) + \tau \, \partial_\tau v(\tau) = u_{GF}(t) + \frac{1}{t}\partial_\tau u_{GF}(t(\tau))$$

$$= u_{GF}(t) - t\partial_t u_{GF}(t) = u_{GF}(t) + tp_{GF}(t)$$

$$= u_{GF}(t_i) - (t - t_i)p_{GF}(t_{i+1}) + tp_{GF}(t_{i+1})$$

$$= u_{GF}(t_i) + t_i p_{GF}(t_{i+1}). \tag{5.48}$$

Thus, we can continue computing

$$\langle r(\tau), \partial_\tau(\tau v(\tau))\rangle = \langle p_{VM}(t), u_i + t_i p_{GF}(t_{i+1})\rangle,$$

$$= J(u_i) + t_i\langle p_{VM}(t), p_{GF}(t_{i+1})\rangle. \tag{5.49}$$

Due to (5.42) and the piecewise constant $p_{GF}(t)$, we have

$$p_{VM}(t) = \frac{1}{t}\left(\sum_{j=1}^{i-1}(t_{j+1} - t_j)p_{GF}(t_{j+1}) + (t - t_i)p_{GF}(t_{i+1})\right).$$

Using the above formula for $p_{VM}(t)$, we can use the (MINSUB) condition to obtain

$$t_i\langle p_{VM}(t), p_{GF}(t_{i+1})\rangle = \frac{t_i}{t}\left\langle\left(\sum_{j=1}^{i-1}(t_{j+1} - t_j)p_{GF}(t_{j+1}) + (t - t_i)p_{GF}(t_{i+1})\right), p_{GF}(t_{i+1})\right\rangle,$$

$$= \frac{t_i}{t}\left(\sum_{j=1}^{i-1}(t_{j+1} - t_j)\langle p_{GF}(t_{j+1}), p_{GF}(t_{i+1})\rangle + (t - t_i)\|p_{GF}(t_{i+1})\|^2\right),$$

$$\overset{\text{(MINSUB)}}{=} t_i\|p_{GF}(t_{i+1})\|^2 \overset{\text{(Theorem 5.8)}}{=} t_i J(p_{GF}(t_{i+1})).$$

By combining the above estimate with (5.49), we obtain

$$J(\partial_\tau(\tau v(\tau))) = J(u(t_i) + t_i p_{GF}(t_{i+1})), \leq J(u(t_i)) + t_i J(p_{GF}(t_{i+1})),$$

$$= \langle r(\tau), \partial_\tau(\tau v(\tau))\rangle,$$

which yields $r(\tau) \in \partial J(\partial_\tau(\tau v(\tau)))$ and hence the assertion.

Conclusion 5 *Let (DDL1) be met. Then, GF, VM, and IF all yield the same spectral representation.*

Proof Theorem 5.6 along with Theorem 5.4 show that $u_{GF}(t) = u_{VM}(t)$, which implies $\phi_{VM}(t) = \phi_{GF}(t)$. Theorem 5.7 tells us that

$$v_{IS}(s) = \partial_s\,(s\,u_{GF}(1/s))$$
$$= u_{GF}(1/s) - \frac{1}{s}\partial_t u_{GF}(1/s).$$

Thus,

$$\tilde{\phi}_{IS}(s) = \partial_s v_{IS}(s) = -\frac{1}{s^2}\partial_t u_{GF}(1/s) - \left(-\frac{1}{s^2}\partial_t u_{GF}(1/s) - \frac{1}{s^3}\partial_{tt} u_{GF}(1/s)\right).$$
$$= \frac{1}{s^3}\partial_{tt} u_{GF}(1/s)$$

The relation $\phi_{IS}(t) = \frac{1}{t^2}\psi_{IS}(1/t)$ of (5.24) now yields

$$\phi_{IS}(t) = t\partial_{tt} u_{GF}(t) = \phi_{GF}(t).$$

5.5.4 Nonlinear Eigendecompositions

As described in the introduction, the eigendecomposition, or, more generally, the singular value decomposition of a linear operator plays a crucial role in the classical filter analysis. Furthermore, we discussed that the notion of eigenvectors has been generalized to an element v_λ with $\|v_\lambda\|_2 = 1$ such there exists a λ with

$$\lambda v_\lambda \in \partial J(v_\lambda).$$

The classical notion of singular vectors (up to a square root of λ) is recovered for a quadratic regularization functional J, i.e., $J(u) = \frac{1}{2}\|Ku\|_2^2$, in which case $\partial J(u) = \{K^*Ku\}$. We are particularly interested in the question in which case our generalized notion of a spectral decomposition admits the classical interpretation of filtering the coefficients of a (nonlinear) eigendecomposition of the input data f.

It is interesting to note that the use of eigendecompositions of linear operators goes beyond purely linear filtering. Denoting an orthonormal transformation matrix by V^T, we can represent the filtered data $\hat{u}$ of f as

$$\hat{u} = VDV^T f =: R(f), \tag{5.50}$$

where D is a diagonal matrix containing the filter as its diagonal.

Popular nonlinear versions of the classical linear filters (5.50) can be obtained by choosing filters adaptively to the magnitude of the coefficients in a new representation. For example, let V be an orthonormal matrix (corresponding to a change of basis). For input data f, one defines $u_{\text{filtered}} = V \, D_{V^T f} \, V^T f$, where $D_{V^T f}$ is a data-dependent diagonal matrix, i.e., $\operatorname{diag}(D_{V^T f}) = g(V^T f)$ for an appropriate function g. Examples for such choices include hard or soft thresholding of the coefficients. In [3], it was shown that these types of soft and hard thresholdings of representation coefficients can be recovered in our framework by choosing $J(u) = \|V^T u\|_1$.

The matrix V arising from the eigendecomposition of a linear operator is orthogonal, i.e., $V V^T = V^T V = Id$. The following theorem shows that significantly less restrictive conditions on the regularization, namely J meeting (DDL1), already guarantee the decomposition of f into a linear combination of generalized eigenvectors.

Theorem 5.8 (Decomposition into eigenfunctions) *Let (DDL1) be met. Then, (up to normalization) the subgradients $p_{GF}(t_{i+1})$ of the gradient flow are eigenvectors of J, i.e., $p_{GF}(t_{i+1}) \in \partial J(p_{GF}(t_{i+1}))$. Hence,*

$$f = P_0(f) + \sum_{i=0}^{N}(t_{i+1} - t_i)p_{GF}(t_{i+1}), \tag{5.51}$$

for N being the index with $u(t_N) = P_0(f)$ is a decomposition of f into eigenvectors of J.

Proof For the sake of simplicity, denote $p_i := p_{GF}(t_i)$ and $u_i := u_{GF}(t_i)$. We already know that $p_{i+1} = K^T q_{i+1}$ for some q_{i+1}, $\|q_{i+1}\|_\infty \le 1$, with $q_{i+1}(l) = \operatorname{sign}(Ku_i)(l)$ for $l \in I_{u_i}$.

$$\langle p_{i+1}, p_{i+1} \rangle = \langle KK^T q_{i+1}, q_{i+1} \rangle,$$
$$= \sum_{l \notin I_{u_{i+1}}} \underbrace{(KK^T q_{i+1})(l)}_{=0,\text{ Proof of 5.6}} \cdot q_{i+1}(l) + \sum_{l \in I_{u_{i+1}}}(KK^T q_{i+1})(l) \cdot q_{i+1}(l).$$

For the second sum, we have

$$(KK^T q_{i+1})(l) = (KK^T)(l, l) \cdot \operatorname{sign}(Ku_i(l)) + \sum_{j \neq i}(KK^T)(l, j) \cdot q_{i+1}(j),$$

$$\overset{\text{Diag. dom. of } KK^T}{\Rightarrow} \operatorname{sign}((KK^T q_{i+1})(l)) = \operatorname{sign}(Ku_i(l)) \text{ or } (KK^T q_{i+1})(l) = 0.$$

Thus, in any case, $l \in I_{u_i}$ and $l \notin I_{u_i}$, we have $(KK^T q_{i+1})(l) \cdot q_{i+1}(l) = |(KK^T q_{i+1})(l)|$, such that

$$\langle p_{i+1}, p_{i+1} \rangle = \langle KK^T q_{i+1}, q_{i+1} \rangle = \sum_{l}|(K \underbrace{K^T q_{i+1}}_{=p_{i+1}})(l)| = \|Kp_{i+1}\|_1,$$

which completes the proof.

Note that Conclusion 5 immediately generalizes this result from the spectral representation arising from the gradient flow to all three possible spectral representations.

It is interesting to see that the subgradients of the gradient flow are (up to normalization) the eigenvectors f is decomposed into. By the definition of $\phi_{GF}(t) = t\partial_{tt}u_{GF}(t) = -t\partial_t p_{GF}(t)$ and the mathematical definition of $\phi_{GF}(t)$ removing the derivative in front of $p_{GF}(t)$, we can see that any filtering approach in the (DDL1) case indeed simply modifies the coefficients of the representation in Theorem 5.8, and therefore establishes a full analogy to linear filtering approaches.

Since any f can be represented as a linear combination of generalized eigenvectors, a conclusion of Theorem 5.8 is the existence of a basis of eigenfunctions for any regularization meeting (DDL1). Note that there can (and in general will), however, be many more eigenfunctions than dimensions of the space, such that the nonlinear spectral decomposition methods cannot be written in a classical setting.

Let us now compare the two representations of f given by Eq. (5.51) and by our usual reconstruction Eq. (5.22), or—since (DDL1) implies we have a polyhedral regularization—the discrete reconstruction given by Eq. (5.41). For simplicity, we assume $P_0(f) = 0$. While one can see that rearranging Eq. (5.41) yields Eq. (5.51), our decomposition of an image into its ϕ parts corresponds to the *change* of the eigenfunctions during the piecewise dynamics. While eigenfunctions of absolutely one-homogeneous regularizations are often highly correlated, their differences can be orthogonal as stated in Theorem 5.5 and therefore nicely separate different scales of the input data.

References

1. G. Gilboa, A spectral approach to total variation, in *SSVM 2013*, Lecture Notes in Computer Science, vol. 7893, ed. by A. Kuijper, et al. (Springer, Berlin, 2013), pp. 36–47
2. G. Gilboa, A total variation spectral framework for scale and texture analysis. SIAM J. Imaging Sci. **7**(4), 1937–1961 (2014)
3. M. Burger, L. Eckardt, G. Gilboa, M. Moeller, Spectral representations of one-homogeneous functionals, in *Scale Space and Variational Methods in Computer Vision* (Springer, Berlin, 2015), pp. 16–27
4. Martin Burger, Guy Gilboa, Michael Moeller, Lina Eckardt, Daniel Cremers, Spectral decompositions using one-homogeneous functionals. SIAM J. Imaging Sci. **9**(3), 1374–1408 (2016)
5. L.R. Rabiner, B. Gold, *Theory and Application of Digital Signal Processing* (Prentice-Hall Inc., Englewood Cliffs, 1975), p. 777
6. M. Benning, M. Burger, Ground states and singular vectors of convex variational regularization methods. Methods Appl. Anal. **20**(4), 295–334 (2013)
7. Y. Meyer, Oscillating patterns in image processing and in some nonlinear evolution equations, March 2001. The 15th Dean Jacquelines B. Lewis Memorial Lectures
8. L.C. Evans, *Partial Differential Equations*, Graduate Studies in Mathematics, vol. 19 (American Mathematical Society, 1991)
9. J. Sporring, J. Weickert, Bounds on shape recognition performance. IEEE T. Image Process **45**(3), 1151–1158 (1999)
10. J-J. Fuchs, Spread representations, in *Asilomar Conference on Signals, Systems, and Computers* (2011)

11. S. Vaiter, G. Peyré, J.M. Fadili, Model consistency of partly smooth regularizers (2014), arXiv:1405.1004
12. V. Duval, G. Peyré, Exact support recovery for sparse spikes deconvolution. Found. Comput. Math. **15**(5), 1315–1355 (2015)
13. T. Brox, M. Welk, G. Steidl, J. Weickert, in *Proceedings of the 4th International Conference on Scale Space Methods in Computer Vision* (Springer, Berlin, 2003), pp. 86–100
14. M.R. Osborne, B. Presnell, B.A. Turlach, A new approach to variable selection in least squares problems. IMA J. Numer. Anal. **20**(3), 389–403 (2000)
15. R.J. Tibshirani, J. Taylor, The solution path of the generalized lasso. Ann. Stat. **39**(3), 1335–1371 (2011)
16. B. Grünbaum, *Convex Polytopes*, Graduate Texts in Mathematics (Springer, Berlin, 2003)
17. M. Moeller, M. Burger, Multiscale methods for polyhedral regularizations. SIAM J. Optim. **23**(3), 1424–1456 (2013)

Chapter 6
Applications Using Nonlinear Spectral Processing

6.1 Generalized Filters

6.1.1 Basic Image Manipulation

Given the decomposition and spectral framework, a rather straightforward application is the attenuation and amplification of certain desired bands. This produces very interesting image effects, of enhancing or attenuating certain textural scales, keeping other image components intact. In Fig. 6.1, a band-pass and its complementary band-stop filters are shown for the bees image. The band of the filter selects spectral components depicted by the red curve in the spectrum, where certain scales appear to be dominant. We see that those are scales of repetitive structures (the beehive), which are decomposed nicely from the input image. In Fig. 6.2, an example of an orange with diminished and enhanced textures is shown. In Fig. 6.3, a certain spatial area is filtered. This is done by a fuzzy mask of the skin area. Attenuating certain bands give softer skin effect, keeping in the very fine textures to preserve a natural look. Note that the basic colors are often in the coarser bands, so the overall coloring maintains natural. In Fig. 6.4, a closeup is shown depicting more clearly the filtering effects.

6.2 Simplification and Denoising

One can use the spectral framework for denoising or for simplifying the image as a preprocessing stage for various tasks such as edge detection, segmentation, or compression. It was shown in [1, Theorem 2.5] that the popular denoising strategy of evolving the scale space flow (5.1) for some time t_1 is equivalent to the particular filter

© Springer International Publishing AG, part of Springer Nature 2018   93
G. Gilboa, *Nonlinear Eigenproblems in Image Processing and Computer Vision*, Advances in Computer Vision and Pattern Recognition,
https://doi.org/10.1007/978-3-319-75847-3_6

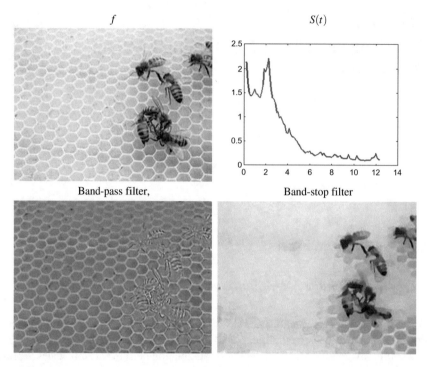

Fig. 6.1 Band-pass filtering example. The honeycomb can be mostly separated from the bees. The band which is filtered is shown in red in the spectrum $S(t)$ (top right)

$$H(t) = \begin{cases} 0 & \text{for } 0 \leq t \leq t_1 \\ \frac{t - t_1}{t} & \text{for } t_1 \leq t \leq \infty \end{cases} \tag{6.1}$$

in the framework of Eq. (5.5). See Fig. Fig. 6.5 for plot of the filter for several time instances. The latter naturally raises the question if this particular choice of filter is optimal for practical applications. While in case of a perfect separation of signal and noise an ideal low-pass filter (5.6) may allow a perfect reconstruction (as illustrated on synthetic data in Fig. 6.6), the spectral decomposition of natural noisy images will often contain noise as well as texture in high-frequency components. Note that the spectral decomposition here is based on inverse-scale-space. In this case t is proportional to λ and noise and fine scale features of high eigenvalues appear on the right side of the spectrum.

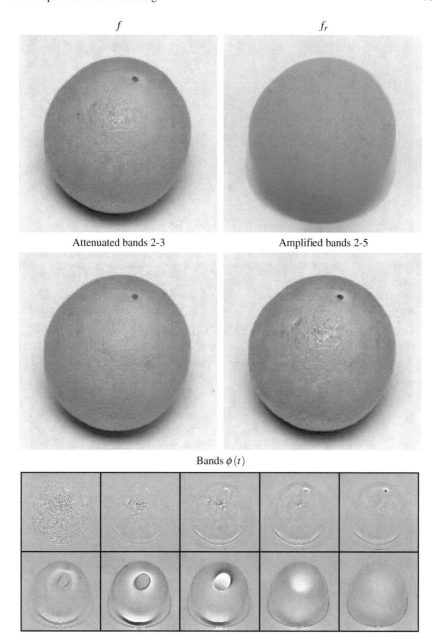

Fig. 6.2 Texture processing. Attenuating or amplifying certain texture layers of an orange. Top—original (left) and residual f_r at time $t = 10$. Middle row—attenuated textures (left) and amplified textures. Bottom row—several examples of texture bands (ϕ functions) are shown: 1, 2, 3, 4, 5, 10, 15, 20, 30, 40

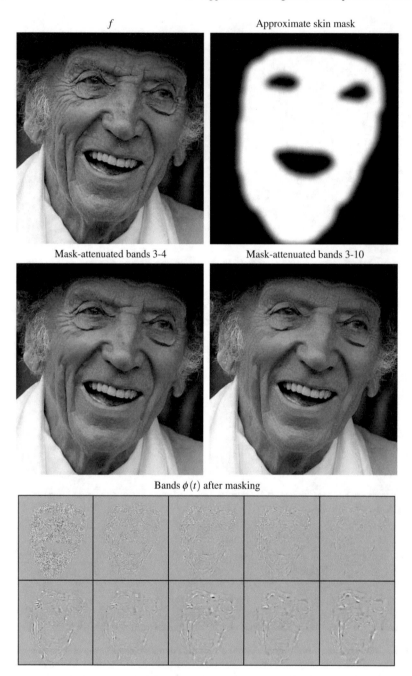

Fig. 6.3 Texture processing. Removing texture layers from a face. Top—original (left) and rough skin mask. Middle row—light texture removal (bands 3, 4 are attenuated, left), strong texture removal (bands 3–10 are attenuated, right). Bottom row—10 first bands (after masking)

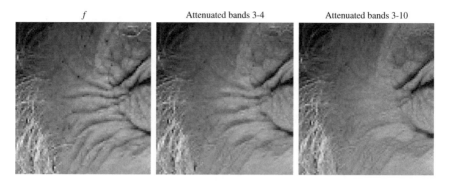

Fig. 6.4 Closeup of Fig. 6.3 (left eye region). From left—original, light texture removal, strong texture removal

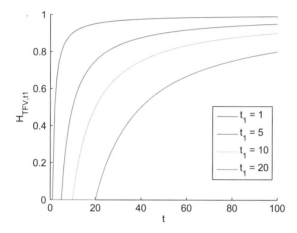

Fig. 6.5 Illustration of Eq. (6.1). TV flow is equivalent to a nonideal low-pass filter in the spectral domain. In the plot, four filters H_{TVF,t_1} are plotted for different values $t_1 = \{1, 5, 10, 20\}$

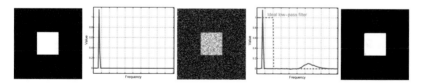

Fig. 6.6 Example for perfect separation of noise and signal via anisotropic TV regularization in the framework of nonlinear spectral decompositions using the inverse scale space flow (therefore, here the direction of the spectrum, with respect to scale, is inverted). From left to right: Clean image, corresponding spectrum of the clean image, noisy image, spectrum of the noisy image with an ideal low-pass filter to separate noise and signal, reconstructed image after applying the ideal low-pass filter

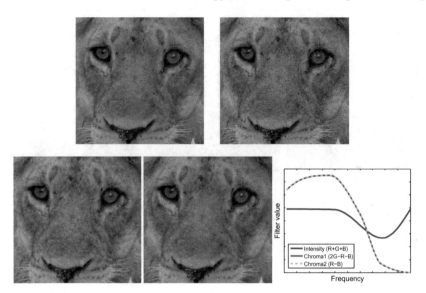

Fig. 6.7 Example for denoising natural images: The original image (upper left) contains a lot of texture. If one tries to denoise the upper right image with an ideal low-pass filter (using TV in a color space that decouples intensities and chromaticities), we obtain the lower left image. Since some of the high frequencies are completely suppressed, the image looks unnatural. A designed filter as shown in the lower right can produce more visually pleasing results (lower middle image). This example was produced using the ISS framework such that the filters are shown in frequency representation. Note that the suppression of color artifacts seems to be significantly more important than the suppression of oscillations

6.2.1 Denoising with Trained Filters

In [2], a first approach to learning optimal denoising filters on a training dataset of natural images demonstrated promising results. In particular, it was shown that optimal filters neither had the shape of ideal low-pass filters nor of the filter arising from evolving the gradient flow. With respect to color, it was concluded that the best filter for each channel should be composed using information from all color channels.

In future research projects, different regularization terms for separating noise and signals in a spectral framework will be investigated. Moreover, the proposed spectral framework allows to filter more general types of noise, i.e., filters are not limited to considering only the highest frequencies as noise. Finally, the complete absence of high-frequency components may lead to an unnatural appearance of the images, such that the inclusion of some (damped) high-frequency components may improve the visual quality despite possibly containing noise. Figure 6.7 supports this conjecture by a preliminary example.

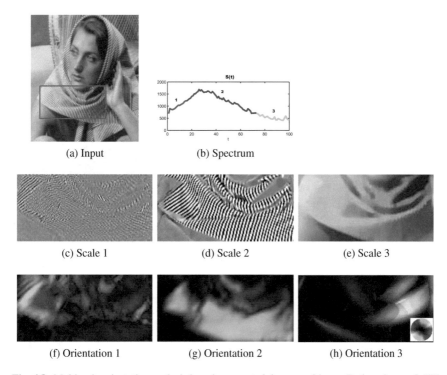

(a) Input   (b) Spectrum

(c) Scale 1   (d) Scale 2   (e) Scale 3

(f) Orientation 1   (g) Orientation 2   (h) Orientation 3

Fig. 6.8 Multiscale orientation analysis based on spectral decomposition. **a** Barbara image, **b** TV spectrum of the image with separated scales marked in different colors. **c–e** Multiscale decomposition, **f–h** Corresponding orientation maps

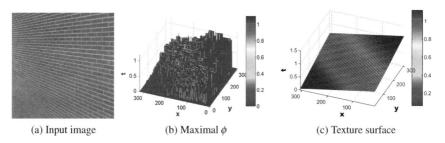

(a) Input image   (b) Maximal ϕ (c) Texture surface

Fig. 6.9 Spectral analysis of a wall image, from [3]. **a** Input image of a brick wall. **b** $S(x) = \max_t \phi(t; x)$. **c** Approximation of $S(x)$ by a plain

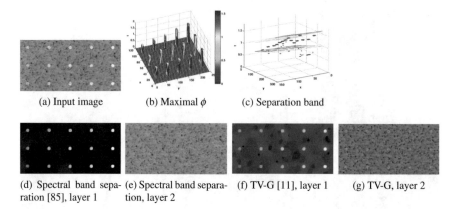

(a) Input image   (b) Maximal ϕ (c) Separation band

(d) Spectral band separation [85], layer 1   (e) Spectral band separation, layer 2   (f) TV-G [11], layer 1   (g) TV-G, layer 2

Fig. 6.10 Decomposition by a separation band in the spectral domain. **a** Input image, **b** The maximal ϕ response, **c** The separation band, **d–e** spectral decomposition, **f–g** TV-G decomposition. Taken from [4]

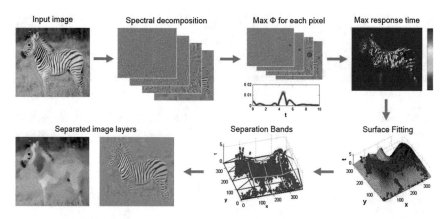

Fig. 6.11 Algorithm flow of the separation surface

6.3   Multiscale and Spatially Varying Filtering Horesh–Gilboa

One can view the spectral representation as an extension to infinite dimensions of multiscale approaches for texture decomposition, such as [6, 7]. In this sense, $\phi(t)$ of (5.3) is an infinitesimal textural element (which goes in a continuous manner from "texture" to "structure", depending on t). A rather straightforward procedure is therefore to analyze the spectrum of an image and either manually or automatically select integration bands that correspond to meaningful textural parts in the image. This was done in [8], where a multiscale orientation descriptor, based on Gabor filters, was constructed. This yields for each pixel a multi-valued orientation field,

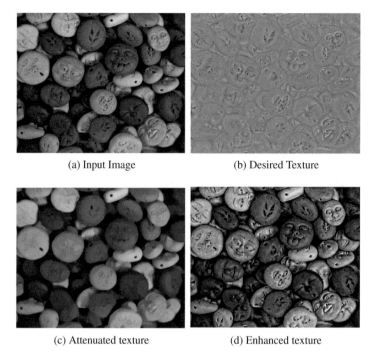

(a) Input Image   (b) Desired Texture

(c) Attenuated texture   (d) Enhanced texture

Fig. 6.12 Example of manipulation of faces' features in the image of faces on stones

which is more informative for analyzing complex textures. See Fig. 6.8 as an example of such a decomposition of the Barbara image.

In [3, 4], a unique way of using the spectral representation was suggested for texture processing. It was shown that textures with gradually varying pattern size, pattern contrast, or illumination can be represented by surfaces in the three-dimensional TV transform domain. A spatially varying texture decomposition amounts to estimating a surface which represents significant maximal response of ϕ, within a time range $[t_1, t_2]$,

$$\max_{t \in [t_1, t_2]} \{\phi(t; x)\} > \varepsilon$$

for each spatial coordinate x. In Fig. 6.11, the main steps of devising an automatic algorithm to find the spatially varying filtering range are shown.

In Fig. 6.9, a wall is shown with gradually varying texture scale and its scale representation in the spectral domain. A decomposition of structures with gradually varying contrast is shown in Fig. 6.10 where the band-spectral decomposition (spatially varying scale separation) is compared to the best TV-G separation of [5] (Fig. 6.11). Manipulating images by enhancing or attenuating certain texture bands can be seen in Figs. 6.12 and 6.13. Image manipulation of the Zebra image (enhancement, color inversion) are shown in Fig. 6.14.

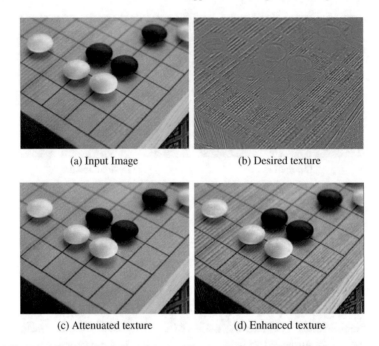

(a) Input Image   (b) Desired texture

(c) Attenuated texture   (d) Enhanced texture

Fig. 6.13 Example of manipulation of the wood texture in the game board image

(a) Input image   (b) Enhanced stripes   (c) Inverted stripes

Fig. 6.14 Stripes extraction and manipulation in the zebra image

6.4 Face Fusion and Style Transfer

In [9], fusion of images was proposed based on the spectral decomposition paradigm. Essentially, different bands could be taken from different images to form new images in a realistic manner. A hard problem is to be able to fuse faces in a realistic manner. The main idea is to first align the main facial features of eyes mouth and nose by nonrigid registration techniques. Then, following decomposition into TV spectral bands, one can mix the fine details of one person into the coarse scale features of another in a seemingly natural way, with minimal artifacts. The concept main steps are illustrated in Fig. 6.15. See two options for face fusion in Fig. 6.16.

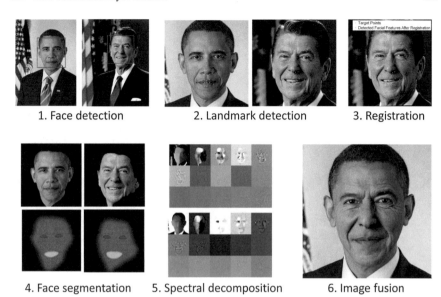

1. Face detection   2. Landmark detection   3. Registration

4. Face segmentation   5. Spectral decomposition   6. Image fusion

Fig. 6.15 Illustration of the pipeline for facial image fusion using nonlinear spectral decompositions

Fig. 6.16 Example of a facial image fusion result obtained by the proposed framework. The left image of Reagan and right image of Obama were used as an input, while the two middle images are synthesized using nonlinear spectral TV filters

Fig. 6.17 Inserting the shark from the left image into the middle image via spectral image fusion yields the image on the right: by keeping low frequencies from the middle image, one obtains highly believable colors in the fusion result. A smooth transition between the inserted object and the background image by a fuzzy segmentation mask (alpha-matting) was used to further avoid fusion artifacts

Fig. 6.18 Example of transforming a photo such that it gives the impression of being an impression-ist painting. Using the spectral decomposition, we extract the brush stroke features of the painting (very left) from high-frequency bands at a certain area (marked in red) and embedded them into the photo image. The result is shown on the right

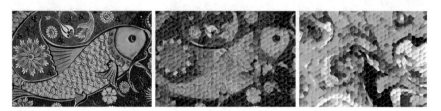

Fig. 6.19 Example of transforming ceramic art such that it gives the impression of being a mosaic. Low frequencies from the ceramic art (left) are extracted and combined with the high frequencies from the mosaic image (right)

One can also fuse objects based on this method (Fig. 6.17) as well as change the perform style transfer of the image. That is, change the material and textural appearance of an image to those of another image, as in Figs. 6.18 and 6.19.

References

1. G. Gilboa, A total variation spectral framework for scale and texture analysis. SIAM J. Imaging Sc. **7**(4), 1937–1961 (2014)
2. M. Moeller, J. Diebold, G. Gilboa, D. Cremers, Learning nonlinear spectral filters for color image reconstruction. In Proceedings of the IEEE International Conference on Computer Vision (ICCV) (2015) p. 289–297
3. D. Horesh, Separation surfaces in the spectral TV domain for texture decomposition, Master thesis, Technion, 2015
4. D. Horesh, G. Gilboa, Separation surfaces in the spectral tv domain for texture decomposition. IEEE Trans. Image Process. **25**(9), 4260–4270 (2016)
5. J.F. Aujol, G. Aubert, L. Blanc-Féraud, A. Chambolle, Image decomposition into a bounded variation component and an oscillating component. JMIV **22**(1), 71–88 (2005)
6. E. Tadmor, S. Nezzar, L. Vese, A multiscale image representation using hierarchical (BV, L2) decompositions. SIAM Multiscale Model. Simul. **2**(4), 554–579 (2004)
7. J. Gilles, Multiscale texture separation. Multiscale Model. Simul. **10**(4), 1409–1427 (2012)
8. D. Horesh, G. Gilboa, Multiscale Texture Orientation Analysis Using Spectral Total-Variation Decomposition, *Scale Space and Variational Methods in Computer Vision* (Springer, Berlin, 2015), pp. 486–497

9. M. Benning, M. Möller, R.Z. Nossek, M. Burger, D. Cremers, G. Gilboa, C.-B. Schönlieb. Nonlinear spectral image fusion. In international conference on scale space and variational methods in computer vision (Springer, Berlin, 2017), p. 41–53

Chapter 7
Numerical Methods for Finding Eigenfunctions

7.1 Linear Methods

Linear eigenvalue problems arise in many fields of science and engineering: in civil engineering, they determine how resistant a bridge is to vibrations; in quantum mechanics, they impose the modes of a quantum system and in fluid mechanics eigenvalue analysis models well the flow of liquids near obstacles. Complex high-dimensional eigenvalue problems arise today in disciplines such as machine learning, statistics, electrical networks, and more. There is vast research and literature, accumulated throughout the years, on numerical solvers for linear eigenvalue problems [1–4]. Given a matrix A, a common practice is to calculate an eigenvalue revealing factorization of A, where the eigenvalues appear as entries within the factors and the eigenvectors are columns in an orthogonal matrix used in the decomposition (e.g., Schur factorization and unitary diagonalization). This is often performed by applying a series of transformations to A in order to introduce zeros in certain matrix entries. This process is done iteratively until convergence. Notable algorithms applying such techniques are the *QR algorithm* [5] or the *divide-and-conquer* algorithm [6]. As a consequence, these methods are appropriate for linear operators on finite-dimensional spaces (matrices), and unfortunately such techniques do not naturally extend to the nonlinear case. However, not all techniques perform a sequence of factorizations (or diagonalization). One of such methods is the *inverse power method* (IPM) and its extension, the *Rayleigh quotient iteration* [2]. Hein and Bühler [7] found a clever way to generalize the Rayleigh quotient iteration to the nonlinear eigenvalue problem case, with the same definition as in (4.2).

© Springer International Publishing AG, part of Springer Nature 2018   107
G. Gilboa, *Nonlinear Eigenproblems in Image Processing and Computer Vision*, Advances in Computer Vision and Pattern Recognition, https://doi.org/10.1007/978-3-319-75847-3_7

7.2   Hein–Buhler

We outline here the method of Hein and Bühler [7]. The authors extend the inverse
power method (for more information on the basic method, see, e.g., [3]) for finding eigenvalues and eigenvectors for matrices to the nonlinear case with one-
homogeneous functionals. In order to understand the method in [7], first let us
consider the Rayleigh quotient that is defined as

$$F_{Rayleigh}(u) = \frac{\langle u, Au \rangle}{\|u\|_2^2}, \tag{7.1}$$

where $A \in \mathbb{R}^{n \times n}$ is a real symmetric matrix and $u \in \mathbb{R}^n$. If u is an eigenfunction of
A, then $F_{Rayleigh}(u) = \lambda$ where λ is the corresponding eigenvalue of u. In [7], the
authors consider functionals F of the form

$$F(u) = \frac{R(u)}{S(u)}, \tag{7.2}$$

where both R and S are convex and $R : \mathbb{R}^n \to \mathbb{R}^+$, $S : \mathbb{R}^n \to \mathbb{R}^+$. One can observe
that the functional in (7.2) is a generalization of the functional in (7.1). A critical
point u^* of F fulfills

$$0 \in \partial R(u^*) - \lambda \partial S(u^*),$$

where ∂R, ∂S are the subdifferentials of R and S, respectively, and $\lambda = \frac{R(u^*)}{S(u^*)}$. We
identify $R(u) = J(u)$ and $S(u) = \frac{1}{2}\|u\|_2^2$. Note that this equation now becomes the
nonlinear eigenvalue problem (4.3).

   The standard (linear) iterative IPM uses the scheme $Au^{k+1} = u^k$ in order to
converge to the smallest eigenvector of A. This scheme can also be represented as
an optimization problem:

$$u^{k+1} = \arg \min_v \frac{1}{2}\langle v, Av \rangle - \langle v, u^k \rangle.$$

This can directly be generalized to the nonlinear case by

$$u^{k+1} = \arg \min_v J(v) - \langle v, u^k \rangle. \tag{7.3}$$

Specifically for one-homogeneous functionals, a slight modification is required and
the minimization problem is given by

$$u^{k+1} = \arg \min_{\|v\| \le 1} J(v) - \lambda^k \langle v, u^k \rangle, \tag{7.4}$$

i.e., adding the constraint that $\|v\| \leq 1$ and the addition of λ^k, where $\lambda^k = \frac{J(u^k)}{\|u^k\|_2^2}$ to the minimization, in order to guarantee descent.

7.3 Nossek–Gilboa

Recently, a flow-based method was suggested in [8] to solve nonlinear eigenvalue problems of one-homogeneous functionals. The idea is to be able to compute numerically as many eigenfunctions as possible, both with low and high eigenvalues. For that a flow-type mechanism, controlled by a time step, which can be very precise, is suggested. It was later realized that several generalizations of the prototype described below can be formulated, solving a variety of nonlinear eigenproblems. A thorough analytical study has shown several settings where such algorithms converge.

Let J be a proper, convex, lower semi-continuous, one-homogeneous functional such that the gradient descent flow (5.1) is well posed. The following flow is considered:

$$u_t = \frac{u}{\|u\|} - \frac{p}{\|p\|}, \quad p \in \partial J(u), \tag{7.5}$$

with $u|_{t=0} = f$, where f admits $\|f\| \neq 0$, $\langle f, 1 \rangle = 0$, $f \in \mathcal{N}(J)^\perp$. The later property can be achieved for any input $\tilde{f}$ by subtracting its projecting onto the nullspace, $f = \tilde{f} - P_0 \tilde{f}$. Thus, we have that $J(f) > 0$. It can easily be shown that under these assumptions $\|u(t)\| \neq 0$ and $\|p(t)\| \neq 0$, $\forall t \geq 0$, so the flow is well defined. We also assume that J is a regularizing functional (e.g., based on derivatives), such as TV or TGV, and it is thus invariant to a global constant change, such that

$$J(u) = J(u + c), \quad \forall u \in \mathcal{X}, c \in \mathbb{R}.$$

Such an assumption may be relaxed, as briefly discussed below.

Remark on existence and uniqueness. The continuous formulation is in our opinion rather elegant and intuitive, thus it is easier to understand the rationale of using such flows. However, at the stage of writing this book, the complete time-continuous theory is still under investigation. Here, the assumption is that the flow exists, for which some general properties are shown in the time-continuous case. A discrete semi-implicit scheme is then proposed which can be computed numerically by iterating standard convex optimization techniques. The numerical scheme is analyzed, where it is shown that many characteristics predicted by the time-continuous formulation are valid in the discrete case as well.

7.3.1 Flow Main Properties

We will now show that this is a smoothing flow in terms of the functional J and an enhancing flow with respect the ℓ^2 norm, where a nontrivial steady state is reached for nonlinear eigenfunctions admitting Eq. (4.3) and only for them.

Theorem 7.1 *Assuming a solution $u(t)$ of the flow (7.5) exists, then it has the following properties:*

Property 1   The mean value of $u(t)$ is preserved throughout the flow:

$$\langle u(t), 1 \rangle = 0.$$

Property 2

$$\frac{d}{dt} J(u(t)) \leq 0,$$

where equality is reached iff u is an eigenfunction (admits (4.3)).
Property 3

$$\frac{d}{dt} \|u(t)\|^2 \geq 0,$$

where equality is reached iff u is an eigenfunction.
Property 4   A necessary condition for steady state $u_t = 0$ holds iff u is an eigenfunction.

Proof   1. From the invariance to constant change, $J(u) = J(u + c)$, using (1.15) it is easy to show that
$$J^*(p) = J^*(p) - \langle c, p \rangle$$

yielding $c\langle p, 1 \rangle = 0$. Let us define $Q(t) = \langle u(t), 1 \rangle$. Using (7.5) and the above equality, we obtain

$$\frac{d}{dt} Q(t) = \langle u_t(t), 1 \rangle = \left\langle \frac{u(t)}{\|u(t)\|} - \frac{p}{\|p\|}, 1 \right\rangle = \frac{1}{\|u(t)\|} \langle u(t), 1 \rangle = \frac{1}{\|u(t)\|} Q(t),$$

with a solution $Q(t) = Be^{\int_0^t \frac{1}{\|u(\tau)\|} d\tau}$, where $B \in \mathbb{R}$. Using the initial condition $u(t = 0) = f$ and the fact that $\langle f, 1 \rangle = 0$ (hence $Q(t = 0) = 0$), yields $B = 0$ resulting in $\langle u(t), 1 \rangle = 0$, $\forall t \geq 0$, i.e., u has mean zero and it is preserved throughout the flow.
   2. For the second claim, we use (4.3) and (4.7) obtaining

$$\frac{d}{dt} J(u(t)) = \langle p, u_t \rangle = \left\langle p, \frac{u}{\|u\|} - \frac{p}{\|p\|} \right\rangle = \frac{J(u)}{\|u\|} - \|p\|.$$

Using (1.21), we conclude $\frac{J(u)}{\|u\|} - \|p\| \le 0$ with equality if and only if p is linearly dependent in u, and hence an eigenfunction.

3. The third claim can be verified in a similar manner by

$$\frac{d}{dt}\left(\frac{1}{2}\|u(t)\|^2\right) = \langle u, u_t \rangle = \left\langle u, \frac{u}{\|u\|} - \frac{p}{\|p\|} \right\rangle = \|u\| - \frac{J(u)}{\|p\|}.$$

4. For the fourth claim, a necessary steady-state condition is

$$u_t = \frac{u}{\|u\|} - \frac{p}{\|p\|} = 0.$$

Therefore, $p = \frac{\|p\|}{\|u\|}u$ and the eigenfunction equation (4.3) holds with $\lambda = \frac{\|p\|}{\|u\|}$. Naturally, on the other direction, if (4.3) holds, $p = \lambda u$, we get $\frac{p}{\|p\|} = \frac{u}{\|u\|}$ and $u_t = 0$.

We note that for Property 1 different arguments can be used, for instance, based on properties of the nullspace of one-homogeneous functionals, as was recently shown in [9].

Notice that from Property 3 of Theorem 7.1 it might seem that $\|u\|_2^2$ can diverge. We show below that as long as the minimal nontrivial eigenvalue (with respect to the regularizer J and the domain) is bounded from below by a positive constant, the 2-norm of u is bounded. Let us first define the minimal nontrivial eigenvalue for a specific value of the regularizer $J(u) = c > 0$, as

$$\lambda_{\min,c} := \min_{u,\, \lambda u \in \partial J(u),\, \lambda > 0,\, J(u) = c} \lambda.$$

Proposition 7.1 *Let the conditions of Theorem 7.1 hold and let $u(t)$ be the solution of the flow of Eq. (7.5) with initial condition f. Then, $\|u(t)\|^2 \le \frac{J(f)}{\lambda_{\min, J(f)}}$, $\forall t \ge 0$.*

Proof We examine the following optimization problem:

$$\max \|u\|^2 \quad \text{s.t. } J(u) = c.$$

To solve this using Lagrange multipliers, we define $\mathcal{L}(u, \alpha) = \|u\|^2 + \alpha(J(u) - c)$, yielding the necessary optimality condition:

$$\frac{\partial \mathcal{L}}{\partial u} = 2u + \alpha p = 0,$$

$$\frac{\partial \mathcal{L}}{\partial \alpha} = J(u) - c = 0.$$

Taking the inner product with respect to u of the first equation (and using $J(u) = \langle u, p \rangle$), we get $\alpha = -\frac{2\|u\|^2}{c}$ where $p = -\frac{2}{\alpha}u$. Thus, the optimal u is an eigenfunction with $\lambda = -\frac{2}{\alpha} = \frac{c}{\|u\|^2} = \lambda_{min,c}$. Moreover, for $c_2 > c_1$, we get $\lambda_{min,c_2} < \lambda_{min,c_1}$. This can be shown by choosing the minimal eigenfunction u_{min,c_1} corresponding to λ_{min,c_1} and multiplying it by c_2/c_1. Then, this is clearly an eigenfunction restricted by $J(u) = c_2$ with a corresponding eigenvalue

$$\lambda_{min,c_2} \leq \lambda|_{J(u)=c_2} = \frac{J(u)}{\|u\|^2} = \frac{c_2}{\|u_{min,c_1}c_2/c_1\|^2} = \frac{c_1}{c_2}\lambda_{min,c_1} < \lambda_{min,c_1}.$$

Using the fact that $c = J(u(t)) \leq J(f)$ yields the bound

$$\|u(t)\|^2|_{J(u(t))=c} \leq \max_{v,\, J(v)=c} \|v\|^2 = \frac{c}{\lambda_{min,c}} \leq \frac{J(f)}{\lambda_{min,J(f)}}, \quad \forall t \geq 0.$$

Remark 1 We remind that $f \in \mathcal{N}(J)^\perp$. It is shown in [10] Lemma 4 that if $p \in \partial J(u)$ then $p \in \mathcal{N}(J)^\perp$. Therefore, since our flow is a linear combination of u and p, we are kept in the subspace $\mathcal{N}(J)^\perp$ and $J(u(t)) > 0, \forall t \geq 0$.

Remark 2 This process usually does not converge to the eigenfunction with the smallest eigenvalue (it depends on the initialization function f). The properties of the flow yield a straightforward bound on λ. As $\|(u(t))\|$ is increasing with time and $J(u(t))$ is decreasing, then when an eigenfunction is reached, its eigenvalue λ is bounded by

$$0 < \lambda = \frac{J(u)}{\|u\|^2} \leq \frac{J(f)}{\|f\|^2}. \tag{7.6}$$

7.3.1.1 Interpretation and Regularity

One can define the ℓ^2 unit vectors in the directions u and p, respectively, as

$$\hat{u} = \frac{u}{\|u\|}, \quad \hat{p} = \frac{p}{\|p\|},$$

with $p \in \partial J(u)$. The flow (7.5) can be rewritten as

$$u_t = \hat{u} - \hat{p}.$$

Thus, there are two competing unit vectors. Notice that for one-homogeneous functionals $\langle u, p \rangle = J(u) > 0$, and therefore the angle between u and p is in the range $(-\frac{1}{2}\pi, \frac{1}{2}\pi)$. Using this observation, we later define an indicator which measures how close a function is to be an eigenfunction, see Sect. 4.5.1. The absolute angle between $\hat{u}$ and $-\hat{p}$ is larger than $\frac{\pi}{2}$, see Fig. 7.1a, where for an eigenfunction $\hat{u}$ and $-\hat{p}$ are exactly at opposite directions (angle π) canceling each others contribution to the flow, enabling a steady-state solution (Fig. 7.1b).

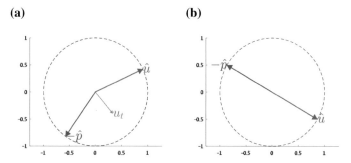

Fig. 7.1 An illustration of the flow and the relation between $\hat{u} = \frac{u}{\|u\|}$ and $\hat{p} = \frac{p}{\|p\|}$. In dashed line, the ℓ^2 unit ball. Figure **a** illustrates the general case where u is not an eigenfunction, while figure **b** illustrates the case where u is an eigenfunction. Note that for this case, $\hat{u}$ and $-\hat{p}$ are exactly opposing one another, yielding $u_t = 0$

Regarding regularity, the flow (7.5) is essentially a time rescale of the gradient flow (5.1) with amplification of u, so as long as there is no blowup in u, the signal becomes smoother in terms of J and regularity is maintained.

7.3.2 Inverse Flow

An alternative flow which works in the inverse direction of (7.5) can also be defined:

$$u_t = -\frac{u}{\|u\|} + \frac{p}{\|p\|}, \quad p \in \partial_u J(u), \tag{7.7}$$

with $u|_{t=0} = f$.

This is an anti-smoothing flow in term of the functional J and a reducing flow with respect to the 2 norm, where also here a necessary steady-state condition is reached for nonlinear eigenfunctions admitting Eq. (4.3) and only for them.

Proposition 7.2 *Assuming a solution $u(t)$ of the flow (7.7) exists, then it has the following properties:*

1.
$$\frac{d}{dt}J(u(t)) \geq 0$$

where equality is reached iff u is an eigenfunction.

2.
$$\frac{d}{dt}\|u(t)\|^2 \leq 0$$

where equality is reached iff u is an eigenfunction.

3. *A necessary condition for steady-state $u_t = 0$ holds iff u is an eigenfunction.*

Proof The proof follows the same lines as the one of Theorem 7.1.

From preliminary experiments, this flow tends to produce non-smooth eigenfunctions with large eigenvalues, as can be expected. We point out this formulation; however, in this book, this direction is not further developed.

7.3.3 Discrete Time Flow

The forward flow, given in (7.5), is discretized in the semi-implicit setting as follows:

$$u^{k+1} = u^k + \Delta t \left(\frac{u^{k+1}}{\|u^k\|} - \frac{p^{k+1}}{\|p^k\|} \right), \quad u^0 = f, \qquad (7.8)$$

with Δt indicating the chosen time step to use. This equation can be reformulated into the following convex optimization problem:

$$u^{k+1} = \operatorname*{arg\,min}_{v} \left\{ J(v) + \frac{\|p^k\|}{2\Delta t} \left(1 - \frac{\Delta t}{\|u^k\|} \right) \left\| \frac{u^k}{1 - \frac{\Delta t}{\|u^k\|}} - v \right\|_2^2 \right\}. \qquad (7.9)$$

Naturally, this minimization is valid (with strictly positive terms and no division by zero) for a time step bounded by

$$\Delta t < \|u^k\|.$$

As we will show hereafter, the 2-norm of u is nondecreasing with iterations; therefore, a sufficient condition for a stable minimization with a constant time step is

$$\Delta t < \|u^0\| = \|f\|. \qquad (7.10)$$

In the following section, we analyze this discrete flow, realized as iterative convex minimizations, and show that many properties align with the time-continuous flow.

7.3.4 Properties of the Discrete Flow

Theorem 7.2 *The solution u^k of the discrete flow of Eq. (7.8), with $\Delta t \le \|u^k\|$, has the following properties:*

Property 1   *The mean value of u^k is zero for any $k = 0, 1, 2, ..,$*

$$\langle u^k, 1 \rangle = 0.$$

Property 2

$$\|p^{k+1}\| \le \|p^k\|.$$

Property 3

$$\|u^{k+1}\| \ge \|u^k\|.$$

Property 4

$$\frac{J(u^{k+1})}{\|u^{k+1}\|} \le \frac{J(u^k)}{\|u^k\|}.$$

Property 5 A sufficient and necessary condition for steady-state $u^{k+1} = u^k$ holds if u^k is an eigenfunction.

Proof Property 1 can be shown by induction, using $\langle f, 1 \rangle = 0$ and $\langle p^k, 1 \rangle = 0$ for all k (as is shown in Theorem 7.1).
Property 2. Let $\alpha = (\|u^k\| - \Delta t)/\|u^k\| > 0$. Equation (7.8) can be written as $\alpha u^{k+1} = u^k - (\Delta t) p^{k+1}/\|p^k\|$. Taking the inner product of p^k with respect to both sides of (7.8) yields

$$\alpha \langle u^{k+1}, p^k \rangle = J(u^k) - \frac{\Delta t}{\|p^k\|} \langle p^{k+1}, p^k \rangle.$$

As $J(u^{k+1}) \ge \langle u^{k+1}, p^k \rangle$, we have the bound

$$\langle p^{k+1}, p^k \rangle \ge (J(u^k) - \alpha J(u^{k+1})) \frac{\|p^k\|}{\Delta t}.$$

Taking the inner product of p^{k+1} with respect to both sides of (7.8) yields

$$\alpha J(u^{k+1}) = \langle u^k, p^{k+1} \rangle - \frac{\Delta t}{\|p^k\|} \|p^{k+1}\|^2.$$

Substituting $\alpha J(u^{k+1})$ in the bound above yields

$$\langle p^{k+1}, p^k \rangle \ge (J(u^k) - \langle u^k, p^{k+1} \rangle) \frac{\|p^k\|}{\Delta t} + \|p^{k+1}\|^2 \ge \|p^{k+1}\|^2.$$

Using $\|p^{k+1} - p^k\|^2 = \|p^{k+1}\|^2 - 2\langle p^{k+1}, p^k \rangle + \|p^k\|^2 \ge 0$, this property is validated.
Property 3. Taking the inner product of each side of the equality (7.8) with respect to itself, we have

$$\|u^{k+1}\|^2 - \|u^k\|^2 = 2\Delta t \left(\frac{\langle u^k, u^{k+1} \rangle}{\|u^k\|} - \frac{\langle u^k, p^{k+1} \rangle}{\|p^k\|} \right) + \Delta t^2 \overbrace{\left\| \frac{u^{k+1}}{\|u^k\|} - \frac{p^{k+1}}{\|p^k\|} \right\|^2}^{\beta}.$$

As $\beta \geq 0$, we check the left term of the right-hand side. Plugging the expression of u^{k+1} as in (7.8) yields

$$
\frac{\langle u^k, u^{k+1} \rangle}{\|u^k\|} - \frac{\langle u^k, p^{k+1} \rangle}{\|p^k\|} = \frac{1}{\|u^k\| - \Delta t} \left(\|u^k\|^2 - \frac{\Delta t \langle u^k, p^{k+1} \rangle}{\|p^k\|} \right) - \frac{\langle u^k, p^{k+1} \rangle}{\|p^k\|}
$$

$$
= \frac{\|u^k\|}{\|u^k\| - \Delta t} \left(\|u^k\| - \frac{\langle u^k, p^{k+1} \rangle}{\|p^k\|} \right).
$$

As $J(u^k) \geq \langle u^k, p^{k+1} \rangle$ (see (1.20)), we get the bound

$$
\frac{\|u^k\|}{\|u^k\| - \Delta t} \left(\|u^k\| - \frac{\langle u^k, p^{k+1} \rangle}{\|p^k\|} \right) \geq \frac{\|u^k\|}{\|u^k\| - \Delta t} \left(\|u^k\| - \frac{J(u^k)}{\|p^k\|} \right) \geq 0,
$$

where for the last inequality we used (1.21) concluding $\|u^{k+1}\|^2 \geq \|u^k\|^2$.
Property 4. We use the fact that u^{k+1} is the minimizer of (7.9), plug in $v = \frac{\|u^{k+1}\|}{\|u^k\|} u^k$ in that expression and use the one-homogeneity property of J to have the bound

$$
J(u^{k+1}) + \frac{\|p^k\|}{2\Delta t} \left(1 - \frac{\Delta t}{\|u^k\|} \right) \overbrace{\left\| \frac{u^k}{1 - \frac{\Delta t}{\|u^k\|}} - u^{k+1} \right\|^2}^{(a)} \leq
$$

$$
\frac{\|u^{k+1}\|}{\|u^k\|} J(u^k) + \frac{\|p^k\|}{2\Delta t} \left(1 - \frac{\Delta t}{\|u^k\|} \right) \overbrace{\left\| \frac{u^k}{1 - \frac{\Delta t}{\|u^k\|}} - \frac{\|u^{k+1}\|}{\|u^k\|} u^k \right\|^2}^{(b)}. \quad (7.11)
$$

For convenience, we denote the norm expressions on the left-hand side and the right-hand side of the inequality, (a) and (b), respectively,

$$
(a) = \frac{\|u^k\|^4}{(\|u^k\| - \Delta t)^2} - 2 \frac{\|u^k\|}{\|u^k\| - \Delta t} \langle u^k, u^{k+1} \rangle + \|u^{k+1}\|^2
$$

$$
(b) = \frac{\|u^k\|^4}{(\|u^k\| - \Delta t)^2} - 2 \frac{\|u^k\|}{\|u^k\| - \Delta t} \|u^k\| \|u^{k+1}\| + \|u^{k+1}\|^2
$$

$$
(a) - (b) = \frac{2\|u^k\|}{\|u^k\| - \Delta t} \left(\|u^k\| \|u^{k+1}\| - \langle u^k, u^{k+1} \rangle \right) \geq 0,
$$

where for the last inequality we used the C-S inequality. Hence, (a) $\geq$ (b). Subtracting the right term on the right-hand side (expression involving (b)) from both sides of (7.11) yields $J(u^{k+1}) \leq \frac{\|u^{k+1}\|}{\|u^k\|} J(u^k)$.
Property 5. Let u^k be an eigenfunction with eigenvalue λ. We have $p^k = \lambda u^k$ and therefore

$$
\lambda = \frac{\|p^k\|}{\|u^k\|}.
$$

For one-homogeneous functionals, we have that also $\tilde{u} = u^k/(1 - \Delta t/\|u^k\|)$ is an eigenfunction with eigenvalue $\tilde{\lambda} = \lambda \cdot (1 - \Delta t/\|u^k\|)$. Hence, (7.9) is in the form of (4.11) with $\alpha = \frac{\|p^k\|}{\Delta t}\left(1 - \frac{\Delta t}{\|u^k\|}\right)$, $f = \tilde{u}$. Its solution is (4.12); therefore,

$$u^{k+1} = (1 - \tilde{\lambda}/\alpha)^+ \frac{u^k}{1 - \frac{\Delta t}{\|u^k\|}} = u^k.$$

The opposite direction is straightforward. If $u^{k+1} = u^k$, Eq. (7.8) yields

$$\frac{u^{k+1}}{\|u^k\|} - \frac{p^{k+1}}{\|p^k\|} = 0,$$

and u^{k+1} is an eigenfunction, $p^{k+1} = \lambda u^{k+1}$, with $\lambda = \frac{\|p^k\|}{\|u^k\|}$.

Remark 3 We should note that although numerically $J(u^k)$ decreases with time (as seen in the Results section), we have not been able to prove this. Alternatively, we have shown in Property 2 the decrease of $\|p^k\|$ (and the decrease of the Rayleigh quotient, see following remark). The remaining properties (1, 3, and 4) are the discrete analog of the properties shown in Theorem 7.1 for the continuous case.

Remark 4 A direct consequence of Theorem 7.2 is the decrease of the Rayleigh quotient. Let the (generalized) Rayleigh quotient at iteration k be defined by $R^k := \frac{J(u^k)}{\|u^k\|^2}$, then Properties 3 and 4 yield that $R^{k+1} \leq R^k$. Also, using an alternative definition $\tilde{R}^k := \frac{\|p^k\|}{\|u^k\|}$ we have from Properties 2 and 3 that $\tilde{R}^{k+1} \leq \tilde{R}^k$. Both definitions coincide in the case of u being an eigenfunction, $p = \lambda u$, where $R = \tilde{R} = \lambda$. As the Rayleigh quotient is decreasing and positive (actually it is bounded from below by the minimal eigenvalue λ_{min}), it converges.

7.3.5 Normalized Flow

We are mostly interested in the spatial structure of eigenfunctions, not their contrast. It is therefore often useful to discuss the normalized case, of eigenfunctions with a unit 2-norm,

$$p = \lambda u, \quad p \in \partial J(u), \quad \|u\|_2 = 1. \tag{7.12}$$

To compute such eigenfunctions, we modify the original flow to include a simple normalization step at each iteration,

$$u^{k+1/2} = u^k + \Delta t \left(\frac{u^{k+1/2}}{\|u^k\|} - \frac{p^{k+1/2}}{\|p^k\|}\right), \quad u^0 = \frac{f}{\|f\|}, \tag{7.13}$$

$$u^{k+1} = \frac{u^{k+1/2}}{\|u^{k+1/2}\|}, \quad p^{k+1} = p^{k+1/2}, \tag{7.14}$$

where (7.13) is computed as in (7.9). Naturally, this iterative procedure does no longer aims at approximating the original time-continuous equation. However, these iterations have many similar properties and specifically their steady state is an eigenfunction. In addition, we get some convenient properties which help us show that the iterations are converging.

For the normalized flow, we can show several properties, which are immediate consequence of its definitions and the characteristics of the original flow. Note that (7.13) is a single iteration in the flow of (7.8), and thus Theorem 7.2 is valid for $u^{k+1/2}$. This leads to the properties stated in the next proposition.

Proposition 7.3 *The normalized iterations flow defined by Eqs. (7.13) and (7.14) has the following properties:*

1. $\langle u^k, 1 \rangle = 0$.
2. $\|u^k\| = 1$, *for all* $k = 0, 1, 2 \ldots$
3. $\|u^{k+1/2}\| \geq 1$.
4. $\|p^{k+1}\| \leq \|p^k\|$.
5. $J(u^{k+1}) \leq J(u^k)$.
6. $J(u^k) \leq \|p^k\|\|u^k\| = \|p^k\|$.
7. *When* u^k *is an eigenfunction with eigenvalue* λ *we have* $J(u^k) = \|p^k\| = \lambda$.
8. *A steady state is reached*, $u_{k+1} = u^k$, *iff* u^k *is an eigenfunction.*

Proof Property 1 follows directly from Theorem 7.2. Property 2 follows from (7.14). Property 3 follows from Property 3 of Theorem 7.2. Property 4 follows from Property 2 of Theorem 7.2. Property 5 follows from Property 4 of Theorem 7.2 and the fact that $\|u^k\| = 1$. Thus, we have $J(u^k) \geq J(u^{k+1/2})/\|u^{k+1/2}\|$. We can now use the fact that J is one-homogeneous and (7.14) to have $J(u^{k+1/2}) = J(u^{k+1})\|u^{k+1/2}\|$. Property 6 is based on Cauchy–Schwartz inequality. For Property 7 in the case of eigenfunction we have $J(u^k) = \langle u^k, p^k \rangle = \langle u^k, \lambda u^k \rangle = \lambda\|u^k\|$ and we use the fact that $\|u^k\| = 1$. The last property follows from Property 5 of Theorem 7.2.

This flow is somewhat easier to analyze, as $\|u^k\|_2 = 1$ by construction and $J(u^{k+1}) \leq J(u^k)$.

7.3.5.1 A Deeper Analysis of the Normalized Flow

We will now show that the iterations of the normalized flow yield $u^{k+1} - u^k \to 0$ in the ℓ^2 sense. This will need a few preparation steps. The general arguments are in the spirit of [11, 12]. However, there are further obstacles in our case, related to additional terms, which are addressed below.

Let us first define the energy

$$F(u, u^k) = J(u) + \frac{\|p^k\|}{2\Delta t}\gamma \|u^k \gamma^{-1} - u\|^2, \qquad (7.15)$$

with $\gamma = 1 - \frac{\Delta t}{\|u^k\|}$. This is the energy which is minimized in (7.9). We define an additional energy which enforces the solution to be within the ℓ^2 unit ball. Let $\chi_{\|\cdot\|\leq 1}$ be the characteristic function of a unit ℓ^2 ball (with values 0 inside the ball and $+\infty$ outside). We define

$$\tilde{F}(u, u^k) = J(u) + \frac{\|p^k\|}{2\Delta t}\gamma \|u^k \gamma^{-1} - u\|^2 + \chi_{\|\cdot\|\leq 1}(u). \qquad (7.16)$$

Lemma 7.1 *The minimizer of (7.16) with respect to the first argument is the solution u^{k+1} of (7.14).*

Proof By rearranging (7.13), we have

$$u^{k+1/2} = \left(u^k - p^{k+1/2}\frac{\Delta t}{\|p^k\|}\right)\gamma^{-1}. \qquad (7.17)$$

Let u^* be the minimizer of $\tilde{F}(u, u^k)$ for some fixed u^k and $p^* \in \partial J(u^*)$. Then, the associated E–L equation satisfies

$$0 \in p^* + \frac{\|p^k\|}{\Delta t}\gamma(u^* - u^k \gamma^{-1}) + \partial\chi_{\|\cdot\|\leq 1}(u^*),$$

and therefore

$$-p^* + \frac{\|p^k\|}{\Delta t}u^k \in \frac{\|p^k\|}{\Delta t}\gamma u^* + \partial\chi_{\|\cdot\|\leq 1}(u^*).$$

We thus have

$$\left(Id + \frac{\Delta t}{\|p^k\|\gamma}\partial\chi_{\|\cdot\|\leq 1}\right)u^* \ni \left(u^k - p^*\frac{\Delta t}{\|p^k\|}\right)\gamma^{-1},$$

and we can deduce that u^* is the projection of $\left(u^k - p^*\frac{\Delta t}{\|p^k\|}\right)\gamma^{-1}$ onto the ℓ^2 unit ball. Comparing this to (7.17), using $\|u^{k+1/2}\| \geq 1$ (Property 3 of Proposition 7.3), we can solve for u^* by

$$u^* = \frac{u^{k+1/2}}{\|u^{k+1/2}\|},$$

which identifies with (7.14).

Lemma 7.2 *For the solution u^{k+1} of (7.14), the following relation hold:*

$$J(u^{k+1}) + \frac{\|p^k\|}{2\Delta t}\gamma\|u^k - u^{k+1}\|^2 \leq J(u^k). \tag{7.18}$$

Proof As $u = u^{k+1}$ is the minimizer of $\tilde{F}(u, u^k)$, we have the relation $\tilde{F}(u^{k+1}, u^k) \leq \tilde{F}(u^k, u^k)$. Since $\|u^k\| = \|u^{k+1}\| = 1$, the last term on the right-hand side of (7.16) is zero and we can write explicitly the bound

$$J(u^{k+1}) + \frac{\|p^k\|}{2\Delta t}\gamma\|u^k\gamma^{-1} - u^{k+1}\|^2 \leq J(u^k) + \frac{\|p^k\|}{2\Delta t}\gamma\|u^k\gamma^{-1} - u^k\|^2.$$

Expanding the norm expression on the left-hand side, using $\|u^k\| = 1$ (which yields also that $\gamma = 1 - \Delta t$) and rearranging leads to

$$J(u^{k+1}) + \frac{\|p^k\|}{2\Delta t}\gamma\|u^k - u^{k+1}\|^2 + E \leq J(u^k),$$

where
$$E = \|p^k\|(1 - \langle u^k, u^{k+1}\rangle) \geq \|p^k\|(1 - \|u^k\|\|u^{k+1}\|) = 0,$$

which yields the assertion of the lemma.

Theorem 7.3 *The sequence of the normalized discrete flow, (7.13) and (7.14), admits* $\lim_{k\to\infty}\|u^{k+1} - u^k\| = 0$.

Proof Since $f \in \mathcal{N}(J)^\perp$ and in a similar manner to Remark 1, we have that $J(u^k)$ is strictly positive. We can define the minimal value that the functional can attain by $J_{min} = \min_{f,k} J(u^k)$. As $J(u^k)$ is decreasing and bounded from below, it converges. From Property 6 of Proposition 7.3, we have $\|p^k\| \geq J(u^k) \geq J_{min}$. Using Lemma 7.2, summing up N iterations of the flow, we have the following bound:

$$\frac{J_{min}}{2\Delta t}\gamma\sum_{k=0}^{N-1}\|u^k - u^{k+1}\|^2 \leq J(u^0) - J(u^N) \leq J(u^0). \tag{7.19}$$

We can now move to the limit.

7.4   Aujol et al. Method

In [12], a generalization of the above flow was suggested. Let $\alpha \in [0; 1]$. The following flow is examined:

$$\begin{cases} u(0) = u_0, \\ u_t = \left(\frac{J(u)}{\|u\|_2^2}\right)^\alpha u - \left(\frac{J(u)}{\|p\|_2^2}\right)^{1-\alpha} p, \qquad p \in \partial J(u). \end{cases} \tag{7.20}$$

Notice that for $\alpha = 1/2$, we retrieve the flow of Nossek and Gilboa (7.5), up to a normalization with $J^{1/2}(u)$.

Proposition 7.4 *For u_0 of zero mean and $\forall \alpha \in [0; 1]$, the trajectory $u(t)$ of the PDE (7.20) satisfies the following properties:*

(i) $\langle u(t), 1 \rangle = 0$.

(ii) $\frac{d}{dt} J(u(t)) \leq 0$ *for almost every t. Moreover, $t \mapsto J(u(t))$ is non increasing. If $\alpha = 0$, we have for almost every t that $\frac{d}{dt} J(u(t)) = 0$ and $t \mapsto J(u(t))$ is constant.*

(iii) $\frac{d}{dt} \|u(t)\|_2 \geq 0$ *and $\frac{d}{dt} \|u(t)\|_2 = 0$ for $\alpha = 1$.*

(iv) *If the flow converge to u^*, we have $p^* = J^{2\alpha-1}(u^*) \frac{\|p^*\|_2^{2(1-\alpha)}}{\|u^*\|_2^{2\alpha}} u^* \in \partial J(u^*)$ so that u^* is an eigenfunction.*

Proof Property (iii) is obtained as follows:

$$
\frac{d}{dt} \frac{1}{2} \|u(t)\|_2^2 = \langle u, u_t \rangle = \left\langle u, \left(\frac{J(u)}{\|u\|_2^2} \right)^{\alpha} u - \left(\frac{J(u)}{\|p\|_2^2} \right)^{1-\alpha} p \right\rangle
$$

$$
= J^{\alpha}(u) \left(\|u\|_2^{2-2\alpha} - \frac{J^{2-2\alpha}(u)}{\|p\|_2^{2-2\alpha}} \right) \geq 0.
$$

For property (ii), we use once again Lemma 3.3 of [13]. For almost every t, it holds

$$
\frac{d}{dt} J(u(t)) = \langle p, u_t \rangle = \left\langle p, \left(\frac{J(u)}{\|u\|_2^2} \right)^{\alpha} u - \left(\frac{J(u)}{\|p\|_2^2} \right)^{1-\alpha} p \right\rangle
$$

$$
= J^{1-\alpha}(u) \left(\frac{J^{2\alpha}(u)}{\|u\|_2^{2\alpha}} - \|p\|_2^{2\alpha} \right) \leq 0.
$$

Since $t \mapsto J(u(t))$ is absolutely continuous (thanks to Lemma 3.3 of [13]), we deduce that it is non increasing.

For the specific case of $\alpha = 1$, a complete theory of the flow is presented, including proof of convergence of the discrete case and existence and uniqueness of the time-continuous case. Further investigation are currently being studied regarding finding callibrable functions, which can be viewed as eigenfunctions with respect to an $L1$ ball instead of $L2$. They admit the following equation

$$
\lambda \operatorname{sign}(u) \in \partial J_{TV}(u).
$$

In a graph context, this is often referred to as the 1-Laplacian eigenproblem.

References

1. J.H. Wilkinson, *The Algebraic Eigenvalue Problem*, vol. 87 (Clarendon Press, Oxford, 1965)
2. L.N. Trefethen, D. Bau III, *Numerical Linear Algebra*, vol. 50. (Siam, 1997)
3. Y. Saad, *Numerical Methods for Large Eigenvalue Problems* (Society for Industrial and Applied Mathematics, 2011)
4. S. Börm, C. Mehl, *Numerical Methods for Eigenvalue Problems* (Walter de Gruyter, New York, 2012)
5. J.G.F. Francis, The QR transformation a unitary analogue to the LR transformation part 1. Comput. J. **4**(3), 265–271 (1961)
6. J.J.M. Cuppen, A divide and conquer method for the symmetric tridiagonal eigenproblem. Numerische Mathematik **36**(2), 177–195 (1980)
7. M. Hein, T. Bühler, An inverse power method for nonlinear eigenproblems with applications in 1-spectral clustering and sparse PCA. In *Advances in Neural Information Processing Systems 23*, ed. by J.D. Lafferty, C.K.I. Williams, J. Shawe-Taylor, R.S. Zemel, A. Culotta (Curran Associates, Inc., 2010), pp. 847–855
8. R. Nossek, G. Gilboa, Flows generating nonlinear eigenfunctions. J. Sci. Comput. (2017) (accepted)
9. M.F. Schmidt, M. Benning, C-B. Schönlieb, Inverse scale space decomposition (2016), arXiv:1612.09203
10. M. Burger, G. Gilboa, M. Moeller, L. Eckardt, D. Cremers, Spectral decompositions using one-homogeneous functionals. SIAM J. Imaging Sci. **9**(3), 1374–1408 (2016)
11. G. Aubert, J.-F. Aujol, A variational approach to removing multiplicative noise. SIAM J. Appl. Math. **68**(4), 925–946 (2008)
12. J-F. Aujol, G. Gilboa, N. Papadakis, Theoretical analysis of flows estimating eigenfunctions of one-homogeneous functionals for segmentation and clustering (2017). HAL Preprint hal-01563922
13. H. Brezis, *Opérateurs maximaux monotones et semi-groupes de contractions dans les espaces de Hilbert* (North Holland, 1973)

Chapter 8
Graph and Nonlocal Framework

8.1 Graph Total Variation Analysis

Let $G = (V, W)$ denote a weighted, undirected symmetric graph, where V is the set of vertices and W a nonnegative similarity matrix. Let $u, v : V \to \mathbb{R}$ functions on the vertices, $p, q : \mathbb{R}^{|V|} \to \mathbb{R}^{|V|}$ operators and $i, j \in V$. We extend the nonlocal mathematical framework in Sect. 2.5.3 to graphs as follows. Denote the inner product for functions:

$$\langle u, v \rangle := \sum_{i \in V} u_i v_i. \tag{8.1}$$

For operators we define a dot product:

$$\langle p, q \rangle := \sum_{ij \in V} p_{ij} q_{ij}. \tag{8.2}$$

The graph gradient $\nabla_G : V \to V \times V$ is defined by

$$(\nabla_G (u))_{ij} := (u_j - u_i)\, w_{ij}. \tag{8.3}$$

We define the graph divergence $div_G : V \times V \to V$ is defined as the adjoint of the graph gradient:

$$(div_G (p))_i := \sum_{j \in V} (p_{ij} - p_{ji})\, w_{ij}, \tag{8.4}$$

and the standard graph Laplacian $\triangle_G : V \to V$ by

$$(\triangle_G (u))_i = \left(\frac{1}{2} div_G (\nabla_G u) \right)_i = \sum_{j \in V} (u(i) - u(j))\, w_{ij} \tag{8.5}$$

© Springer International Publishing AG, part of Springer Nature 2018
G. Gilboa, *Nonlinear Eigenproblems in Image Processing and Computer Vision*, Advances in Computer Vision and Pattern Recognition,
https://doi.org/10.1007/978-3-319-75847-3_8

Furthermore, we define the graph total variation $J_{TV}(u) : \mathbb{R}^{|V|} \to \mathbb{R}$ by

$$J_{TV}(u) = \|\nabla_G(u)\|_1 = \frac{1}{2} \sum_{ij \in V} \left| u^i - u^j \right| w_{ij}. \tag{8.6}$$

This is a convex, absolutely one-homogeneous functional. Note that the graph operators differ from the nonlocal operators in Sect. 2.5.3 by the power of the weights in the definitions, i.e., w_{ij} instead of $\sqrt{w_{ij}}$. We can now restate properties (2.28)–(2.31) in a graph notation:

"Divergence theorem":

$$\sum_{i \in V} (\mathrm{div}_G(p))_i = 0. \tag{8.7}$$

The Laplacian is self-adjoint

$$\langle \Delta_G u, u \rangle = \langle u, \Delta_G u \rangle \tag{8.8}$$

and negative semidefinite

$$\langle \Delta_G u, u \rangle = -\langle \nabla_G u, \nabla_G u \rangle \leq 0. \tag{8.9}$$

The standard graph Laplacian also induces the quadratic form

$$\langle \Delta_G u, u \rangle = \frac{1}{2} \sum_{ij \in V} w_{ij} \left(u_i - u_j \right)^2. \tag{8.10}$$

We will see how these definitions relate to the analysis of the nonlinear graph operators throughout this chapter in a natural way.

8.2   Graph P-Laplacian Operators

Spectral analysis of the graph p-Laplacian has shown great interest due to the relationship between its second eigenfunction and the Cheeger cut which will be introduced later in this chapter. We define the graph p-Laplacian $\Delta_p(u) : \mathbb{R}^{|V|} \to \mathbb{R}^{|V|}$ in a coordinate form by

$$\left(\Delta_p u \right)_i = \sum_{j \in V} w_{ij} \phi_p \left(u(i) - u(j) \right), \quad i \in V, \tag{8.11}$$

where

$$\phi_p(x) = |x|^{p-1} \, sign(x) . \tag{8.12}$$

This is a nonlinear generalization of the standard graph Laplacian, which was shown [1] to generalize the quadratic form for $p \geq 1$ by

$$\langle u, \triangle_p u \rangle = \frac{1}{2} \sum_{ij \in V} w_{ij} \left| u_i - u_j \right|^p. \tag{8.13}$$

We call u a p-eigenfunction if there exist a constant λ such that

$$\triangle_p u = \lambda \phi_p (u), \tag{8.14}$$

where λ is the corresponding p-eigenvalue. Note that this p-eigenproblem formulation corresponds to $p > 1$. We will later see the difference for the case of $p = 1$.

Theorem 8.1 (Hein et al. 2009) *The function u is a p-eigenfunction if and only if it is a critical point of the following functional:*

$$F_p (u) := \frac{\langle u, \triangle_p u \rangle}{\|u\|_p^p} \tag{8.15}$$

and the corresponding p-eigenvalue is given by $\lambda = F_p (u)$.

Corollary 8.1 *For any $\alpha \neq 0$ the function αu is also a p-eigenfunction corresponding to the same p-eigenvalue, i.e., all p-eigenfunctions are scale invariant.*

This theorem allows us to calculate and order p-eigenfunctions by their corresponding functional value, however, using this characteristic we can only search for p-eigenfunctions independently. This means that a minimization process might reach the same p-eigenfunction over again, e.g., the second p-eigenfunction, which is the global minimum of $F_p (u)$. The next theorem gives a more complete characteristic of the p-Laplacian eigenfunctions by formulating an optimization problem where several different p-eigenfunctions form a local solution.

Theorem 8.2 (Luo et al. 2010) *A p-eigenfunction u has the following properties.*

Property 1 *If u is nonconstant p-eigenfunction then*

$$\sum_{i \in V} \phi_p (u_i) = 0. \tag{8.16}$$

Property 2 *If u,v are p-eigenfunctions associated with different p-eigenvalues then*

$$\sum_{i \in V} \phi_p (u) \phi_p (v) \approx 0. \tag{8.17}$$

Property 3 *If $u_1, \ldots, u_n$ are n p-eigenfunctions associated with unique p-eigenvalues $\lambda_1, \ldots, \lambda_n$, then $u_1, \ldots, u_n$ are local solution to the following problem:*

$$min_{u_1,\ldots,u_n} \sum_{i=1}^{n} F_p(u_i). \qquad (8.18)$$

$$s.t \sum_{i \in V} \phi_p(u_k(i))\,\phi_p(u_l(i)) \approx 0, \quad \forall k \neq l. \qquad (8.19)$$

Note that this theorem holds for $p \geq 1$. This problem is intractable, however, in (Luo et al. 2010) they proposed a gradient decent approximation algorithm, and proved that it converges and satisfies the orthogonality constraint.

8.3   The Cheeger Cut

The idea behind the ratio cut criteria is to partition the graph such that the similarity between the partitions is low and both parts are of relatively similar size. Here are interested in the Cheeger cut (CC) which has proved to outperform previous graph cuts such as the ratio cut and normalized cut, and is defined by

$$CC\left(A, \overline{A}\right) = \frac{cut\left(A, \overline{A}\right)}{min\left\{|A|, |\overline{A}|\right\}}, \qquad (8.20)$$

where

$$cut\left(A, \overline{A}\right) = \sum_{i \in A,\, j \in \overline{A}} w_{ij}. \qquad (8.21)$$

Minimizing the CC is NP-hard and so to solve it several relaxations were proposed, using its relation to the total variation, e.g., [2–5], where for an indicator function $u = \chi_A$ we have that

$$J_{TV}(u) = cut\left(A, \overline{A}\right). \qquad (8.22)$$

Now, to derive an equivalent to the CC we note that

$$\|\chi_A - median(\chi_A)\|_1 = min\left(|A|, |\overline{A}|\right) \qquad (8.23)$$

and so we rewrite the CC by

$$CC\left(A, \overline{A}\right) = \frac{J_{TV}(\chi_A)}{\|\chi_A - median(\chi_A)\|_1} = \frac{J_{TV}(u)}{\|u - median(u)\|_1}. \qquad (8.24)$$

The following theorem characterizes the CC minimization as a total variation minimization problem which turns out to be good practical relaxation.

Theorem 8.3 [2] *Consider the problem*

$$\lambda = min_u \ \frac{J_{TV}\,(u)}{\|u - median\,(u)\|_1}.$$ (8.25)

There is a binary valued minimizer, and

$$\lambda = min_A\ CC\left(A, \overline{A}\right).$$ (8.26)

Furthermore, ∃t s.t the indicator $\chi_{u>t}$ is also a minimizer.

An immediate corollary from this theorem using 8.15 further connects the CC problem to the second 1-eigenfunction.

Corollary 8.2 *The optimal CC is derived by thresholding the second 1-eigenfunction.*

Proof Equation 8.25 can be written as

$$\lambda = min_{median(u)=0}\ F_1\,(u)\,.$$ (8.27)

From (8.15) we know that the second 1-eigenfunction is the solution to this minimization which from (8.1) is known to have zero median.

8.4 The Graph 1-Laplacian

The graph 1-Laplacian eigenfunction analysis and its relation to the CC were introduced in a computational point of view by (Hein 2010), and in a more rigorous mathematical view by [6, 7]. The graph 1-Laplacian is given by

$$(\triangle_1 u)_i = \sum_{j \in V} w_{ij}\ \text{sign}\left(u_i - u_j\right),\ \ i \in V,$$ (8.28)

here it is important to consider the full set valued sign function:

$$(\text{sgn}\,(x))_i = \begin{cases} 1 & x_i > 0 \\ -1 & x_i < 0 \\ [-1, 1] & x_i = 0 \end{cases}$$ (8.29)

which directly implies that the 1-Laplacian is a set-valued function as well. This fact makes the 1-Laplacian a special case of a p-Laplacian operator which is very hard to analyze. This difficulty can be seen by the following definition of the 1-eigenproblem. A function u is called a 1-eigenfunction if it holds

$$0 \in \triangle_1 (u) - \lambda \operatorname{sgn} (u) , \tag{8.30}$$

where $\lambda \in \mathbb{R}$ is the corresponding 1-eigenvalue of u. Theorem 8.1 then states:

Theorem 8.4 *u is a 1-eigenfunction if and only if it is a critical point of*

$$F_1 (u) := \frac{\langle u, \triangle_1 u \rangle}{\|u\|_1} = \frac{J_{TV} (u)}{\|u\|_1} \tag{8.31}$$

and the corresponding 1-eigenvalue is given by $\lambda = F_1 (u)$.

Note that here a critical point corresponds to the non-differentiable case. The following two characteristics of 1-eigenfunctions are not necessarily true for p-eigenfunctions in general.

Proposition 8.1 (Hein 2010) *Any non-constant 1-eigenfunction has median zero.*

Proposition 8.2 *If $u \neq 0$ is a 1-eigenfunction associated with the 1-eigenvalue λ, then $\operatorname{sign} (u)$ is a 1-eigenfunction associated with the same 1-eigenvalue λ.*

Proof By definition

$$0 \in \sum_j w_j \operatorname{sign} \left(u_i - u_j \right) - \lambda sign (u_i) . \tag{8.32}$$

If $sign (u_i) = sign (u_j)$ we have

$$\operatorname{sign} \left(u_i - u_j \right) \subseteq \operatorname{sign} \left(\operatorname{sign} (u_i) - \operatorname{sign} (u_j) \right) = \operatorname{sign} (0) = [-1, 1] \tag{8.33}$$

otherwise $sign (u_i) \neq sign (u_j)$, then

$$\operatorname{sign} \left(u_i - u_j \right) = \operatorname{sign} \left(\operatorname{sign} (u_i) - \operatorname{sign} (u_j) \right) \tag{8.34}$$

Furthermore, $\forall x$ we have $\operatorname{sign} (\operatorname{sign} (x)) = \operatorname{sign} (x)$ and so

$$0 \in \sum_{ij} w_{ij} \operatorname{sign} \left(\operatorname{sign} (u_i) - \operatorname{sign} (u_j) \right) - \lambda \operatorname{sgn} (\operatorname{sign} (u_i)) \tag{8.35}$$

and $\operatorname{sign} (u)$ is a 1-eigenfunction with the corresponding 1-eigenvalue λ. This theorem allows us the derive bi-valued 1-eigenfunctions by calculating any nonzero 1-eigenfunctions. It also shows that thresholding nonzero 1-eigenfunctions are invariant under thresholding at zero.

8.5 The *p*-flow

Generalized flows in the spirit of Nossek–Gilboa were proposed by [8], with a comprehensive theoretical analysis and a proof of convergence. However, such flows do not converge to the double-nonlinear eigenvalue problem of (8.30). In this work, we propose a new flow we term *p-flow* which can solve (8.30) and is presented next. We define the p-flow of u by the diffusion equation

$$u_t = \frac{\phi_p}{\langle u, \phi_p \rangle} - \frac{\triangle_p}{\langle u, \triangle_p \rangle}, \quad u(t = 0) = u_0 \tag{8.36}$$

The p-flow is well defined for $p > 1$, however, it is not the case for $p = 1$. As we have seen above, the sign function and the graph 1-Laplacian are the subgradient set-valued functions $\partial \|u\|_1$ and $\partial J_{TV}(u)$ relatively. Thus, we define the 1-flow by

$$u_t \in \frac{\partial \|u\|_1}{\|u\|_1} - \frac{\partial J_{TV}(u)}{J_{TV}(u)}, \quad u(t = 0) = u_0. \tag{8.37}$$

Our main interest here is the graph 1-Laplacian and so We will further concentrate on 1-flow. However, this analysis can be easily generalized to the simpler case of the p-flow where $p > 1$. For the rest of this chapter, we denote by $p = \triangle_p(u)$ the 1-Laplacian and by $q = sign(u)$ the set-valued sign function. With this notation, we can write the 1-flow by

$$u_t \in \frac{q}{\langle u, q \rangle} - \frac{p}{\langle u, p \rangle}. \tag{8.38}$$

8.5.1 Flow Main Properties

Theorem 8.5 *The solution u^k of the 1-flow has the following properties:*

Property 1   *The flow reaches steady state only if u is a 1-eigenfunction.*
Property 2   *The L^2 norm of u is preserved throughout the flow.*
Property 3   *The flow satisfies the subgradient decent equation.*

$$u_t \in -\partial_u \log (F_1(u)) \tag{8.39}$$

Proof   1. If the 1-flow reaches a steady state than

$$0 \in \frac{q}{\langle u, q \rangle} - \frac{p}{\langle u, p \rangle} \tag{8.40}$$

which is equivalent to the definition of a 1-eigenfunction for u

$$0 \in p - \frac{\langle u, p \rangle}{\langle u, q \rangle} q = p - F_1(u) q. \tag{8.41}$$

2. We can simply verify this by

$$\frac{d}{dt}\left(\frac{1}{2}\langle u, u \rangle\right) = \langle u, u_t \rangle = \left\langle u, \frac{q}{\langle u, q \rangle} - \frac{p}{\langle u, p \rangle}\right\rangle = \frac{\langle u, q \rangle}{\langle u, q \rangle} - \frac{\langle u, p \rangle}{\langle u, p \rangle} = 0. \tag{8.42}$$

3. First, we note that

$$\log(F_1(u)) = \log\left(\frac{J_{TV}(u)}{\|u\|_1}\right) = \log(J_{TV}(u)) - \log(\|u\|_1). \tag{8.43}$$

then,

$$u_t = \frac{q}{\langle u, q \rangle} - \frac{p}{\langle u, p \rangle} \in \frac{\partial \|u\|_1}{\|u\|_1} - \frac{\partial J_{TV}(u)}{J_{TV}(u)} = -\partial_u \log(F_1(u)). \tag{8.44}$$

8.5.2 Numerical Scheme

We consider the following semi-explicit scheme to our flow:

$$\frac{u_{k+1} - u_k}{\delta t} \in \frac{q_k}{\langle u_k, q_k \rangle} - \frac{p_{k+1}}{\langle u_k, p_k \rangle}. \tag{8.45}$$

Proposition 8.3 *The numerical scheme admits:*

1. $F_1(u_{k+1}) \le F_1(u_k)$.
2. $\|u_{k+1}\|_2^2 \ge \langle u_{k+1}, u_k \rangle \ge \|u_k\|_2^2$.

Proof First, notice that we have, for all $\delta t > 0$:

$$u_{k+1} = \underset{u}{argmin}\, H(u) := \frac{\|u - u_k\|_2^2}{2\delta t} - \frac{\langle u, q_k \rangle}{\|u_k\|_1} + \frac{J_{TV}(u)}{J_{TV}(u_k)} \tag{8.46}$$

1. One has

$$H(u_{k+1}) \le H(u_k) = 0, \tag{8.47}$$

$$\frac{\|u_{k+1} - u_k\|_2^2}{2\delta t} - \frac{\langle u_{k+1}, q_k \rangle}{\langle u_k, q_k \rangle} + \frac{\langle u_{k+1}, p_{k+1} \rangle}{\langle u_k, p_k \rangle} \le 0, \tag{8.48}$$

$$\frac{\langle u_{k+1}, p_{k+1} \rangle}{\langle u_k, p_k \rangle} \le \frac{\langle u_{k+1}, q_k \rangle}{\langle u_k, q_k \rangle} \le \frac{\langle u_{k+1}, q_{k+1} \rangle}{\langle u_k, q_k \rangle}, \tag{8.49}$$

$$\frac{\langle u_{k+1}, p_{k+1}\rangle}{\langle u_{k+1}, q_{k+1}\rangle} \leq \frac{\langle u_k, p_k\rangle}{\langle u_k, q_k\rangle}. \tag{8.50}$$

2. One has $u_{k+1} = u_k + \delta t \left(\frac{q_k}{\langle u_k, q_k\rangle} - \frac{p_{k+1}}{\langle u_k, p_k\rangle}\right)$. Taking the scalar product of u_{k+1} with itself:

$$\langle u_{k+1}, u_{k+1}\rangle = \langle u_{k+1}, u_k\rangle + \delta t \left(\frac{\langle u_{k+1}, q_k\rangle}{\langle u_k, q_k\rangle} - \frac{\langle u_{k+1}, p_{k+1}\rangle}{\langle u_k, p_k\rangle}\right), \tag{8.51}$$

$$\langle u_{k+1}, u_{k+1}\rangle \geq \langle u_{k+1}, u_k\rangle. \tag{8.52}$$

On the other hand, we have

$$\langle u_k, u_k\rangle = \langle u_{k+1}, u_k\rangle - \delta t \left(\frac{\langle u_{k+1}, q_k\rangle}{\langle u_k, q_k\rangle} - \frac{\langle u_{k+1}, p_{k+1}\rangle}{\langle u_k, p_k\rangle}\right), \tag{8.53}$$

$$\langle u_k, u_k\rangle \leq \langle u_{k+1}, u_k\rangle. \tag{8.54}$$

8.5.3 Algorithm

We implement our 1-flow using the prime dual algorithm. For a fixed u_k we define

$$G(u) = J_{TV}(u_k)\left(\frac{\|u - u_k\|_2^2}{2\delta t} - \frac{\langle u, q_k\rangle}{\|u_k\|_1}\right), \quad F = \|\cdot\|_1, \quad K = \nabla_G. \tag{8.55}$$

The proximal operator of G is calculated

$$prox_{\tau G}\left(u'\right) = argmin_u \left\{ J_{TV}(u_k)\left(\frac{\|u - u_k\|_2^2}{2\delta t} - \frac{\langle u, q_k\rangle}{\|u_k\|_1}\right) + \frac{\|u - u'\|_2^2}{2\tau}\right\}. \tag{8.56}$$

To solve it we take the derivative of the argument equals to zero and get

$$J_{TV}(u_k)\left(\frac{u - u_k}{\delta t} - \frac{q_k}{\|u_k\|_1}\right) + \frac{u - u'}{\tau} = 0, \tag{8.57}$$

$$u = \frac{\tau J_{TV}(u_k)\delta t}{\tau J_{TV}(u_k) + \delta t}\left(\frac{q_k}{\|u_k\|_1} + \frac{u_k}{\delta t} + \frac{u'}{\tau J_{TV}(u_k)}\right). \tag{8.58}$$

and so we can write the proximal operator in explicit coordinate form by

$$\left(prox_{\tau G}\left(u'\right)\right)_i = \frac{\tau J_{TV}(u_k)\delta t}{\tau J_{TV}(u_k) + \delta t}\left(\frac{q_k}{\|u_k\|_1} + \frac{u_k}{\delta t} + \frac{u'}{\tau J_{TV}(u_k)}\right)_i. \tag{8.59}$$

The proximal operator of $F = \|\cdot\|_1$ is known to be

$$(prox_{\sigma F^*}(p))_{ij} = \frac{p_{ij}}{max\left\{1, |p_{ij}|\right\}}. \tag{8.60}$$

The adjoint operator of K is given by $K^*(p) = -div_G(p)$, and its norm is given by

$$max_{\|u\|=1}\left\{\|K(u)\|\right\} = max_{\|u\|=1}\left\{\|K^*(u)\|\right\} = max_j\left\{\left(\sum_i (w_{ij})^2\right)^{\frac{1}{2}}\right\}, \tag{8.61}$$

which completes the preliminaries for the prime dual algorithm.

Algorithm 1 1-flow prime-dual

Input: Non-negative Similarity matrix W and current function state u_k.

Output: Next function state $u_{k+1} = argmin_u \frac{\|u-u_k\|_2^2}{2\delta t} - \frac{\langle u,q_k\rangle}{\|u_k\|_1} + \frac{J_{TV}(u)}{J_{TV}(u_k)}$.

1. Initialize u^0, p^0, $\bar{u}^0 = u^0$ $tau = sigma = 0.99/\|K\|$, $\theta \approx 1$, $dt > 0$, $\varepsilon > 0$.
2. **repeat**
3. $p^{n+1} = prox_{\sigma F^*}\left(p^n + \sigma K\left(\bar{u}^n\right)\right)$
4. $u^{n+1} = prox_{\tau G}\left(u^n - \tau K^*\left(p^{n+1}\right)\right)$
5. $\bar{u}^{n+1} = u^{n+1} + \theta\left(u^{n+1} - u^n\right)$
6. **until** $\|u^{n+1} - u^n\| < \varepsilon$
7. **return** u^{n+1}

References

1. S. Amghibech, Eigenvalues of the discrete p-Laplacian for graphs. Ars Combinatoria **67**, 283–302 (2003)
2. X. Bresson, T. Laurent, D. Uminsky, J. Von Brecht, Multiclass total variation clustering in *Advances in Neural Information Processing Systems*. 1421–1429 (2013)
3. M. Hein, T. Bühler, An inverse power method for nonlinear eigenproblems with applications in 1-spectral clustering and sparse PCA. In *Advances in Neural Information Processing Systems* (2010)
4. D. Luo, H. Huang, C. Ding, F. Nie, On the eigenvectors of p-Laplacian. Mach. Learn. **81**(1), 37–51 (2010)
5. A. Szlam, X. Bresson. Total Variation, Cheeger Cuts. ICML. (2010)
6. K. Chang, S. Shao, Zhang, Cheeger's cut, maxcut and the spectral theory of 1-Laplacian on graphs, D. Sci. China Math. **60**, 1963 (2017). https://doi.org/10.1007/s11425-017-9096-6
7. K.C. Chang, Spectrum of the 1-Laplacian and Cheeger's Constant on Graphs. J. Graph. Theo. **81**(2), 167–207 (2016)
8. J-F. Aujol, G. Gilboa, N. Papadakis, Theoretical analysis of flows estimating eigenfunctions of one-homogeneous functionals for segmentation and clustering (2017). HAL Preprint hal-01563922

Chapter 9
Beyond Convex Analysis—Decompositions with Nonlinear Flows

9.1 General Decomposition Based on Nonlinear Denoisers

Assume we have a "coarsening" operator $\mathscr{T}$ that suppresses delicate structures within images. For example, $\mathscr{T}$ can be convolution with a Gaussian kernel, total variation smoothing, BM3D denoising [2], etc. To obtain multiple abstractions of an image f, we can repeatedly apply $\mathscr{T}$ so as to create a scale space like flow

$$
\begin{aligned}
\boldsymbol{u}_0 &= \boldsymbol{f}, \\
\boldsymbol{u}_n &= \mathscr{T}(\boldsymbol{u}_{n-1}), \quad n = 1 \ldots T - 1.
\end{aligned}
\tag{9.1}
$$

This process gradually removes larger and larger structures from the image, while retaining only the geometries that are best preserved by the operator.

Our goal is to use the flow (9.1) for decomposing images into layers that capture different scales. A naive way to do so is to compute differences between consecutive time steps along the flow, namely

$$
\begin{aligned}
\boldsymbol{\phi}_n^{\text{naive}} &= \boldsymbol{u}_n - \boldsymbol{u}_{n+1}, \quad n = 0 \ldots T - 2, \\
\boldsymbol{\phi}_{T-1}^{\text{naive}} &= \boldsymbol{u}_{T-1}.
\end{aligned}
\tag{9.2}
$$

This strategy was adopted in various works for structure–texture separation and for filtering of details at specific scales (using coarsening operators based on L_0 smoothing [3], bilateral filtering [4], guided filtering [5], etc.).

Unfortunately, the decomposition (9.2) *lacks scale specificity*. That is, each layer in (9.2) typically mixes structures with quite a large range of scales. To make this statement more precise, let us first define the term "scale", which often takes on different meanings. Recall that in the Fourier transform, a single scale (a delta function in the Fourier domain) corresponds to a sine function with a particular frequency. In the TV transform [6], on the other hand, a single scale captures convex smooth sets with a specific perimeter-to-area ratio (e.g., disks with a particular radius).

© Springer International Publishing AG, part of Springer Nature 2018
G. Gilboa, *Nonlinear Eigenproblems in Image Processing and Computer Vision*, Advances in Computer Vision and Pattern Recognition,
https://doi.org/10.1007/978-3-319-75847-3_9

Then how can we define scale for general decompositions? Note that in both examples, delta functions in the transform domain corresponds to signals that are invariant under the associated operator. In the Fourier transform the associated operator is convolution with some kernel (sines are invariant under convolution), and in the TV decomposition, the associated operator is TV smoothing (e.g., disks are invariant under TV smoothing).

The naive decomposition (9.2) lacks this property. Namely, a signal f which is invariant under the operator $\mathscr{T}$ (looks the same along the flow up to contrast changes), does not appear at a single scale in the decomposition.

9.1.1   A Spectral Transform

We can observe that scale is actually related to the *rate of decay of details throughout the flow*. For example, in the case of a linear scale space constructed by Gaussian smoothing, sines with different frequencies decay at different rates. Similarly, in the case of the TV flow, disks with different radii decay at different rates. This suggests that to obtain a spectral decomposition, we need to seek a representation of the flow (9.1) in terms of layers that decay at different rates.

$$
u_n = \sum_{\tau=1}^{K} h_\tau(n)\, \phi_\tau,
\tag{9.3}
$$

where ϕ_τ is the layer associated with scale τ and $h_\tau(n)$ is the decay profile of that layer. In this representation, the layers $\{\phi_\tau\}$ constitute the decomposition of the image f, from which the flow was generated.

Generally, we expect the decay profiles $\{h_\tau(n)\}$ to depend only on the operator $\mathscr{T}$ and not on the input image f. Let us assume for the moment that these profiles are known and are linearly independent. Then, using (9.3), we can solve for the layers $\{\phi_\tau\}$ given the flow $\{u_n\}$. Thus, a general framework for obtaining spectral decompositions w.r.t. arbitrary operators is: (i) construct the flow (9.1); (ii) solve for the layers in (9.3).

Assuming we order the decays from fast to slow, the first layers (small τ) would capture small-scale details, whereas the last layers (large τ) would capture large scale details. Thus, τ can be thought of as scale, and $1/\tau$ can be thought of frequency.

9.1.2   Inverse Transform, Spectrum, and Filtering

Without loss of generality, we assume that $h_\tau(0) = 1$ for every τ. Now, since the first image in the flow u_0 equals the input image f, we have from (9.3) that

$$f = \sum_{\tau=1}^{K} \boldsymbol{\phi}_{\tau}. \tag{9.4}$$

Namely, the image f can be represented as a sum of detail layers at different scales. We can thus think of (9.4) as an inverse transform.

It is insightful to compare (9.4) with classical decomposition methods. Specifically, in linear transforms, each layer is composed of a linear combination of predefined functions. For example in 2D Fourier, $\boldsymbol{\phi}_{\tau}$ is a linear combination of 2D sines with frequencies that satisfy $\omega_x^2 + \omega_y^2 = 1/\tau^2$. In wavelets, the layer $\boldsymbol{\phi}_{\tau}$ is a linear combination of translations of a scaled version of the mother wavelet, $\psi(x/\tau)$. In our case, on the other hand, the layers can generally be *content-dependent* and are not necessarily confined to any linear subspace.

The representation (9.4) allows us to construct a spectral plot of the image, so as to analyze which scales are dominant. Following [6], we can define the spectrum of f as

$$S(\tau) = \|\boldsymbol{\phi}_{\tau}\|_1. \tag{9.5}$$

An alternative definition, proposed in [7], is

$$S(\tau) = \sqrt{\langle f, \boldsymbol{\phi}_{\tau}\rangle}. \tag{9.6}$$

This latter construction admits a Parseval identity (see Sect. 9.3) similarly to the Fourier transform, which may be desirable in certain analyses.

The representation (9.4) also allows to perform spectral filtering, by attenuating or enhancing certain scales in the reconstruction. Specifically, given a filter $G(\tau)$ in the scale domain, we can construct a filtered version of f as

$$\tilde{f} = \sum_{\tau} \boldsymbol{\phi}_{\tau}\, G(\tau). \tag{9.7}$$

In particular, similarly to [6], we can define high-pass (HP), low-pass (LP), band-pass (BP), and band-stop (BS) filters as

$$G_{\mathrm{HP}}(\tau) = \begin{cases} 1 & \tau \le \tau_{\mathrm{c}}, \\ 0 & \text{else}, \end{cases} \quad G_{\mathrm{BP}}(\tau) = \begin{cases} 1 & \tau_1 \le \tau \le \tau_2, \\ 0 & \text{else}, \end{cases}$$

$$G_{\mathrm{LP}}(\tau) = 1 - G_{\mathrm{HP}}(\tau), \quad G_{\mathrm{BS}}(\tau) = 1 - G_{\mathrm{BP}}(\tau).$$

9.1.3 Determining the Decay Profiles

Our discussion so far assumed that we know the decay profiles $\{h_{\tau}(n)\}$ associated with the operator $\mathscr{T}$. In certain simple cases, we can derive analytical expressions for

those decays. For example, in a Gaussian-smoothing flow, the decays are exponential
and (9.3) reduces to the Fourier transform (see Sect. 9.3). In the TV flow, the decays
are linear and (9.3) reduces to the spectral TV transform [6] (see Sect. 9.3). However,
for a general operator $\mathscr{T}$, it is typically impossible to derive closed-form expressions
for $\{h_\tau(n)\}$.

To overcome this difficulty, we propose to learn the decay profiles. In fact, we
show that under mild assumptions, this can be done from even just a single image.
Thus, in effect, we can *blindly* decompose any image w.r.t. any operator $\mathscr{T}$ (without
assuming that the associated decays are known a priori).

We make the following two assumptions, which seem to be reasonable for most
"coarsening" operators:

A1   *Scale invariance*: We assume that scaling in the spatial domain leads to trans-
      lation of the spectrum in the scale domain. For this property to hold true, all
      the decay profiles must be time-scaled versions of a single prototype decay $v(t)$.
      Namely,

$$h_\tau(n) = v(n/\tau), \tag{9.8}$$

where $v : \mathbb{R}^+ \to \mathbb{R}^+$ satisfies $v(0) = 1$. Note that this is the case in both the
Gaussian-smoothing flow and the TV flow.

A2   *Sparsity*: We assume that in natural images, each pixel is associated with only
      a small number of dominant scales (e.g., the scale of the object it belongs to and
      the scale of the texture on that object). This implies that each location in the image
      is active (nonzero) in only a small number of the layers $\{\boldsymbol{\phi}_\tau\}$.

As we show next, these two postulations significantly simplify the decay estimation
task.

9.2   Blind Spectral Decomposition

Equation (9.3) can be written in matrix form as

$$\boldsymbol{U} = \boldsymbol{H}\boldsymbol{\Phi}, \tag{9.9}$$

where the rows of $\boldsymbol{U}$ are the flow images $\{\boldsymbol{u}_n^T\}$, the rows of $\boldsymbol{\Phi}$ are the decomposition
layers $\{\boldsymbol{\phi}_\tau^T\}$, and the columns of $\boldsymbol{H}$ are the decays $\{h_\tau\}$. Therefore, the blind decomposition problem can be stated as follows: Given the flow $\boldsymbol{U}$ constructed from the
input image, determine both the decomposition $\boldsymbol{\Phi}$ and the decay profiles $\boldsymbol{H}$.

Obviously, problem (9.9) is highly ill-posed. However, Assumptions A1 and A2
provide very strong constraints on $\boldsymbol{\Phi}$ and $\boldsymbol{H}$, which significantly reduce the number
of degrees of freedom. Specifically, Assumption A1 implies that the columns of

H are all time-scaled versions of one single prototype decay vector v (the discrete analog of the function $v(t)$ in (9.8)). We can thus write

$$H = \left(D^{(1)}v \; D^{(2)}v \cdots D^{(K)}v \right), \tag{9.10}$$

where the matrix $D^{(\tau)}$ performs downscaling by a factor of K/τ (we use linear interpolation). Moreover, Assumption A2 implies that the columns of Φ are sparse (each pixel in the image appears in only a small number of layers). Therefore, to recover Φ, we need to minimize

$$\min_{v, \Phi} \| U - H(v)\,\Phi \|_{\mathrm{F}}^2, \tag{9.11}$$

over all prototype vectors v and all column-wise sparse matrices Φ. We note that F stands for the Frobenius norm.

Note that (9.11) is similar to the problem of dictionary learning, where a set of signals (the columns of U) are to be sparsely represented (with coefficients Φ) over a dictionary H. We thus adopt the common paradigm in dictionary learning [8, 9], and solve this problem iteratively, by alternating between updating H and updating Φ. However, there are two fundamental distinctions from the classical dictionary learning setting, which prevent us from using existing approaches "as is".

First, as mentioned above, in our case the dictionary H depends on one single unknown vector v, which affects the H-update step. Second, due to its special structure, this dictionary is very coherent (i.e., the correlation between the columns of H can be close to 1). This is because two decays at slightly different speed are very correlated atoms. Thus, the Φ-update step cannot be performed with popular sparse coding methods such as orthogonal matching pursuit (OMP) [10], because those techniques fail when the dictionary is coherent [11] (see the [1] for elaborated explanation and illustrations).

Updating H. Fixing the matrix Φ, Problem (9.11) becomes a quadratic program w.r.t. H. Furthermore, since H depends linearly on v, this is, in fact, a quadratic program w.r.t.@ v. The solution to this problem is given by $v^* = R^{-1}b$, where $b = \sum_k D^{(k)T} U \phi_k$, $R = \sum_{k,k'} \phi_k^T \phi_{k'} D^{(k)T} D^{(k')}$.

Updating Φ. Fixing the matrix H, Problem (9.11) becomes separable and the update of Φ can be done for each column independently. Thus, using superscripts to denote columns, we obtain a set of problems (one for each pixel) of the form

$$\min_{u^\ell} \| U^\ell - H\Phi^\ell \|_{\mathrm{F}}^2, \tag{9.12}$$

where the minimization is over all sparse vectors u^ℓ. Here, U^ℓ is the time profile of the ℓth pixel along the flow, and Φ^ℓ is the corresponding vector of coefficients that determine which decays (columns of H) participate in its representation. Fortunately, in our setting, H has a very unique structure which simplifies the support recovery task. Based on this structure, we suggest a new sparse coding algorithm called *decay matching pursuit* (DMP).

Algorithm 2 Decay Matching Pursuit (DMP)

Input: Time profile $u(n)$, dictionary of decays $\{h_j(n)\}_{j=1}^K$ with vanishing times $\{T_j\}_{j=1}^K$.
Output: Selected atoms $\Omega = \{j_m\}$ and corresponding coefficients $\{\phi_{j_m}\}$.

   Initialize iteration number $m \leftarrow 1$, residual $r(n) \leftarrow u(n)$, current atom index $i \leftarrow K$, set of selected atoms $\Omega = \emptyset$.

   repeat

      Add current atom: $\Omega \leftarrow \Omega \cup i$.

      Calculate $c_{i,j}$ and $e_{i,j}$ according to (9.13), for every $j \in [1, \ldots, i)$.

      Find: $j^* \leftarrow \mathrm{argmin}_{j \in [1,i)} e_{i,j}$.

      Set representation coefficient $\phi_m \leftarrow c_{i,j^*}$

      Update residual: $r(n) \leftarrow r(n) - \phi_i h_i(n)$.

      Update current atom: $i \leftarrow j^*$.

      $m \leftarrow m + 1$.

   until $i = 1$ or $e < e_T$

As initialization for the decays, a linear decay profile could be used (other options like choosing the decay of a random pixel also worked).

DMP is a greedy algorithm, similarly to OMP. However, as opposed to OMP, it takes advantage of the fact that an atom (a column of $\boldsymbol{H}$), which decays to zero at time T_i, has no contribution to the signal representation for $n \geq T_i$. Therefore, in DMP, the atom selection process starts from the slowest decay atom and sequentially proceeds to the fastest decay atom. In each iteration, we select the atom that minimizes the norm of the residual of the representation obtained so far.

We first choose the slowest atom and define the residual to be $r(n) = u(n)$. Then, at each iteration, assuming the last atom we chose was $h_i(n)$, we compute the normalized representation error $e_{i,j}$ obtained with this atom over the interval $[T_j, T_i]$ for every candidate $T_j < T_i$. Namely,

$$e_{i,j} = \frac{1}{T_i - T_j} \sum_{n=T_j}^{T_i} (u(n) - c_{i,j} h_i(n))^2, \tag{9.13}$$

where the coefficient $c_{i,j}$ is chosen to minimize $e_{i,j}$. That is, $c_{i,j} = \sum_{n=T_j}^{T_i} u(n) h_i(n) / \sum_{n=T_j}^{T_i} h_i^2(n)$.

9.3 Theoretical Analysis

We briefly present here the main theoretical results regarding the proposed framework. To keep a concise presentation, we refer the readers to [1] for proofs and details.

9.3.1 Generalized Eigenvectors

In the Fourier case, as well as the TV transform, an eigenvector, with respect to the generating operator (Laplacian, or 1-Laplacian, respectively), translates to a delta in the transform domain. We would like to show that this property is valid also in the general case, under quite general assumptions.

Let us define a generalized eigenvalue problem associated with a (possibly) non-linear operator $\mathcal{T}$ by

$$\mathcal{T}(\boldsymbol{u}) = \lambda \boldsymbol{u}, \quad \lambda \in \mathbb{R} \tag{9.14}$$

We assume that for any eigenvector, multiplication by a constant yields an eigenvector (with a possibly different eigenvalue). That is, if $\boldsymbol{u}$ admits (9.14), then

$$\mathcal{T}(c\boldsymbol{u}) = k_c \lambda c \boldsymbol{u}, \quad , \forall c \in \mathbb{R}, \tag{9.15}$$

for some $k_c \in \mathcal{R}$. For example, in the linear case $k_c = 1$, and for the 1-Laplacian $k_c = 1/c$. The above assumption implies that an eigenvector   only changes its magnitude throughout the flow, so that a single decay profile captures the complete flow behavior, as we formalize next.

Theorem 9.1 *Let $\mathcal{T}$ admit (9.15) and f admit (9.14). Then*

1. *An optimal decay dictionary exists which is 1-sparse.*
2. *Under this dictionary, the representation $\boldsymbol{\Phi}$ of f contains a single nonzero row.*

Spectrum. The spectrum (9.6) satisfies $\sum_\tau S^2(\tau) = \mathbf{1}^T \boldsymbol{\Phi} f$. Therefore, it admits a Parseval-type equality

$$\|f\|_2^2 = f^T f = (\mathbf{1}^T \boldsymbol{\Phi}) f = \|S\|_2^2.$$

9.3.2 Relation to Known Transforms

The Fourier transform and the TV transform can be viewed as special cases within our framework, based on linear diffusion and TV flow, respectively. To simplify the presentation we work in the continuous domain.

Fourier Transform. Let the input image f be expressed by the inverse Fourier transform $f(x) = \int_\omega F(\omega) e^{j\omega x} d\omega$, where F is the Fourier transform of f. Let g_t be a Gaussian with $\sigma = \sqrt{2t}$ and G_t its Fourier transform, $G_t(\omega) = c e^{-t\omega^2}$. Then the flow $u(t)$ can be defined by linear diffusion, or in an analog manner, by $u(t) = g_t * f = c \int_\omega F(\omega) e^{-t\omega^2} e^{j\omega x} d\omega$, where $*$ denotes convolution. With exponentially decaying functions of the form $H(\tau, t) = e^{-t\tau}$, the entire flow can be represented by $u(t) = \int_0^\infty \phi(\tau) H(\tau, t) d\tau$, hence $\phi(\tau) \propto F(\tau)$.

TV Transform. The TV transform is defined in [6] and is based on gradient descent with respect to the total variation energy (TV flow). Let $u(t)$ be the TV flow solution with $u(t = 0) = f$, where $t \in [0, \infty)$ is the flow's time parameter. The TV transform is defined as

$$\phi(t) = u_{tt}t, \tag{9.16}$$

where u_{tt} is the second time derivative of u. It is known that eigenfunctions with respect to the TV-(sub)-gradient operator (1-Laplacian) yield a linear decay under this flow. In our framework, when the coarsening flow is TV flow, having linearly decaying prototype v, we reach an exact analog to the TV transform.

Theorem 9.2 *Let $u(t)$ be a solution of the TV flow. Let $\{H(\tau, t)\}_{\tau \in \mathbb{R}_+}$ be a family of functions defined by $H(\tau, t) = \max\left\{1 - \frac{t}{\tau}, 0\right\}$. Then $u(t) = \int_0^\infty \phi(\tau) H(\tau, t) d\tau.$, where $\phi(t)$ is defined by (9.16).*

This can be shown by expressing $u(t)$ as above and differentiating twice with respect to t.

References

1. O. Katzir. On the scale-space of filters and their applications, Master thesis, Technion, 2017
2. K. Dabov, A. Foi, V. Katkovnik, K. Egiazarian, Image denoising by sparse 3-d transform-domain collaborative filtering. IEEE Trans. Image Process. **16**(8), 2080–2095 (2007)
3. L. Xu, C. Lu, Y. Xu, J. Jia, Image smoothing via L0 gradient minimization, in *ACM Transactions on Graphics (TOG)*, vol. 30 (ACM, 2011), p. 174
4. C. Tomasi, R. Manduchi, Bilateral filtering for gray and color images, in *ICCV '98* (1998), pp. 839–846
5. K. He, J. Sun, X. Tang, Guided image filtering, in *European Conference on Computer Vision* (Springer, 2010), pp. 1–14
6. G. Gilboa, A total variation spectral framework for scale and texture analysis. SIAM J. Imaging Sci. **7**(4), 1937–1961 (2014)
7. M. Burger, G. Gilboa, M. Moeller, L. Eckardt, D. Cremers, Spectral decompositions using one-homogeneous functionals. SIAM J. Imaging Sci. **9**(3), 1374–1408 (2016)
8. K. Engan, S.O. Aase, J.H. Husoy, Method of optimal directions for frame design, in *1999 IEEE International Conference on Acoustics, Speech, and Signal Processing, 1999. Proceedings*, vol. 5 (IEEE, 1999), pp. 2443–2446
9. M. Aharon, M. Elad, A. Bruckstein, *rmk*-svd: An algorithm for designing overcomplete dictionaries for sparse representation. IEEE Trans. Signal Process. **54**(11), 4311–4322 (2006)
10. Y.C. Pati, R. Rezaiifar, P.S. Krishnaprasad, Orthogonal matching pursuit: Recursive function approximation with applications to wavelet decomposition, in *1993 Conference Record of The Twenty-Seventh Asilomar Conference on Signals, Systems and Computers, 1993* (IEEE, 1993), pp. 40–44
11. J.A. Tropp, A.C. Gilbert, Signal recovery from random measurements via orthogonal matching pursuit. IEEE Trans. Inf. Theory **53**(12), 4655–4666 (2007)

Chapter 10
Relations to Other Decomposition Methods

10.1 Decomposition into Eigenfunctions

Let us consider the discrete case of J being a semi-norm on $\mathbb{R}^n$. One of the fundamental interpretations of linear spectral decompositions arises from it being the coefficients for representing the input data in a new basis. This basis is composed of the eigenvectors of the transform. For example, in the case of the cosine transform, the input signal is represented by a linear combination of cosines with increasing frequencies. The corresponding spectral decomposition consists of the coefficients in this representation, which therefore admit an immediate interpretation.

Although the proposed nonlinear spectral decomposition does not immediately correspond to a change of basis anymore, it is interesting to see that the property of representing the input data as a linear combination of (generalized) eigenfunctions can be preserved: In [1] it was shown that for the case of $J(u) = \|Vu\|_1$ and V being any orthonormal matrix, the solution of the scale space flow (5.1) meets

$$Vu(t) = \text{sign}(\zeta) \ \max(|\zeta| - t, 0), \tag{10.1}$$

where $\zeta = Vf$ are the coefficients for representing f in the orthonormal basis of V. It is interesting to see that the subgradient $p(t)$ in (5.1) admits a componentwise representation of $Vp(t)$ as

$$(Vp)_i(t) = \begin{cases} \text{sign}(\zeta_i) & \text{if } |\zeta_i| \geq t, \\ 0 & \text{else.} \end{cases} \tag{10.2}$$

The latter shows that $p(t)$ can be represented by $V^T q(t)$ for some $q(t)$ which actually meets $q(t) \in \partial \|Vp(t)\|_1$. In a single equation, this means $p(t) \in \partial J(p(t))$ and shows that the $p(t)$ arising from the scale space flow are eigenfunctions. Integrating the scale space flow Eq. (5.1) from zero to infinity and using that there are only finitely many times at which $p(t)$ changes, one can see that one can indeed represent f as a linear

© Springer International Publishing AG, part of Springer Nature 2018
G. Gilboa, *Nonlinear Eigenproblems in Image Processing and Computer Vision*, Advances in Computer Vision and Pattern Recognition,
https://doi.org/10.1007/978-3-319-75847-3_10

combination of eigenfunctions. Our spectral decomposition $\phi(t)$ then corresponds to the *change* of the eigenfunctions during the (piecewise) dynamics of the flow.

The case of $J(u) = \|Vu\|_1$ for an orthogonal matrix V is quite specific. It essentially recovers the linear spectral analysis exactly. We have shown in Chap. 5 that the property of finitely many times at which $p(t)$ changes can be proved for any polyhedral one-homogeneous regularization. If in addition the subdifferentials are sufficiently regular, e.g., as in the case of $J(u) = \|Vu\|_1$ and VV^* being diagonally dominant, then the property of $p(t)$ being an eigenfunction for all times t remains true.

10.2   Wavelets and Hard Thresholding

As we have seen above, the gradient flow with respect to regularizations of the form $J(u) = \|Vu\|_1$ has a closed-form solution for any orthogonal matrix V. From Eq. (10.2) one deduces that

$$\phi(t) = \sum_i \zeta_i \delta(t - |\zeta_i|) v_i,$$

where v_i are the rows of the matrix V, and $\zeta_i = (Vf)_i$. In particular, using definition (5.13) we obtain

$$(S_3(t))^2 = \sum_i (\zeta_i)^2 \delta(t - |\zeta_i|).$$

Peaks in the spectrum therefore occur ordered by the magnitude of the coefficients $\zeta_i = (Vf)_i$: The smaller $|\zeta_i|$ the earlier it appears in the wavelength representation ϕ. Hence, applying an ideal low-pass filter (5.6) on the spectral representation with cutoff wavelength t_c is exactly the same as hard thresholding by t_c, i.e., setting all coefficients ζ_i with magnitude less than t_c to zero.

10.2.1   Haar Wavelets

Wavelet functions are usually continuous and cannot represent discontinuities very well. A special case is the Haar wavelets which are discontinuous and thus are expected to represent much better discontinuous signals. The relation between Haar wavelets and TV regularization in one dimension has been investigated, most notably in [2]. In higher dimensions, there is no straightforward analogy, in the case of isotropic TV, as it is not separable, contrary to the wavelet decomposition (thus one gets disk-like shapes as eigenfunctions as oppose to rectangles).

We will show here that even in the one-dimensional case TV eigenfunctions can represent signals in a much more concise manner.

Let $\psi_H(x)$ be a Haar mother wavelet function defined by

$$\psi_H(x) = \begin{cases} 1 & \text{for } x \in [0, \frac{1}{2}), \\ -1 & \text{for } x \in [\frac{1}{2}, 1), \\ 0 & \text{otherwise.} \end{cases} \tag{10.3}$$

For any integer pair $n, k \in \mathbb{Z}$, a Haar function $\psi_H^{n,k}(x) \in \mathbb{R}$ is defined by

$$\psi_H^{n,k}(x) = 2^{n/2}\psi_H(2^n x - k). \tag{10.4}$$

We now draw the straightforward relation between Haar wavelets and TV eigenfunctions.

Proposition 10.1 *A Haar wavelet function $\psi_H^{n,k}$ is an eigenfunction of TV with eigenvalue $\lambda = 2^{(2+n/2)}$.*

Proof One can express the mother wavelet as

$$\psi_H(x) = B_{\frac{1}{2}}(x) - B_{\frac{1}{2}}\left(x - \frac{1}{2}\right),$$

where B_w is a unit peak of width w, as defined in (4.16). In general, any Haar function can be defined as

$$\psi_H^{n,k}(x) = h^n\left(B_{w^n}(x - x_0^{n,k}) - B_{w^n}(x - x_0^{n,k} - w^n)\right),$$

with $h^n = 2^{(n/2)}$, $w^n = 2^{-(n+1)}$ and $x_0^{n,k} = 2^{-n}k$. Thus, based on Proposition 4.1, we have that for any $n, k \in \mathbb{Z}$, $\psi_H^{n,k}$ is an eigenfunction of TV with $\lambda = \frac{2}{h^n w^n} = 2^{(2+n/2)}$.

In Figs. 10.1, 10.2 and 10.3 we show the decomposition of a signal f composed of three peaks of different widths. The TV spectral decomposition shows five numerical deltas (corresponding to the five elements depicted in Fig. 10.2). On the other hand, the Haar wavelet decomposition needs 15 elements to represent this signal, thus the representation is less sparse. In the 2D case, Fig. 10.4, we see the consequence of an inadequate representation, which is less adapted to the data. Haar thresholding is compared to ideal TV low-pass filtering, where blockly artifacts are clearly observed in the Haar case.

10.3 Rayleigh Quotients and SVD Decomposition

Here, we show that the known relation in the linear case of eigenfunctions as extrema of the Rayleigh quotient also holds in the nonlinear case. There some deeper discussion of the possibility of finding an orthogonal decomposition based on nonlinear

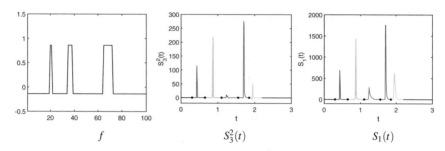

Fig. 10.1 Example for a spectral decomposition: We evaluate the flow of Eq. (5.1) for given data f shown in the left. We compute the spectral decomposition (5.3) and compute a (discretized) version of the power spectral S_3^2 and S_1 from (5.13) and (5.10) shown in the middle and right plot, respectively. Since there appear five distinct peaks, one can integrate the spectral ϕ components at the respective time intervals (=band-pass filter), visualized in different colors, to obtain the five components shown in Fig. 10.2

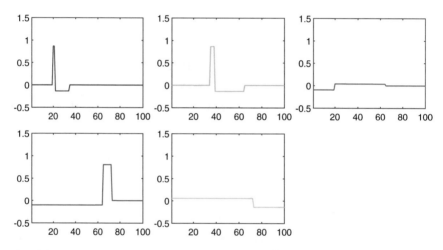

Fig. 10.2 Decomposing f (Fig. 10.1 left) into the five elements represented by the peaks in the spectrum using spectral TV

eigenfunctions and a result by Benning–Burger [3] showing cases where this cannot happen. However, this result, which may seem pessimistic at first, does not contradict our spectral decomposition framework. Our theoretical analysis has shown that the ϕ are orthogonal, but they are differences of eigenfunctions in the general case (they can be eigenfunctions in some specific cases, as 0 is also an eigenfunction).

In the classical Hilbert space case, the Rayleigh quotient is defined by

$$R_M(v) := \frac{v^T M v}{v^T v}, \tag{10.5}$$

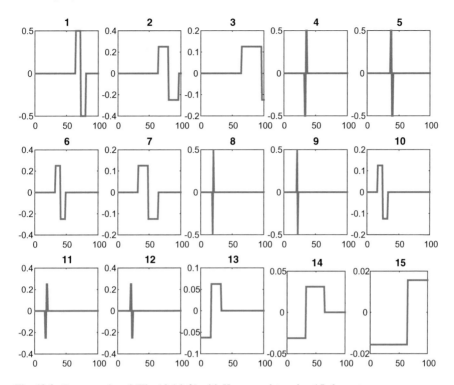

Fig. 10.3 Decomposing f (Fig. 10.1 left) with Haar wavelets using 15 elements

Fig. 10.4 Comparison of wavelet Haar hard thresholding to spectral TV hard thresholding (ideal LPF). Although both representations can handle well discontinuities, the spectral TV representation is better adapted to the image edges and produces few artifacts

where for the real-valued case M is a symmetric matrix and v is a nonzero vector. It can be shown that, for a given M, the Rayleigh quotient reaches its minimum value at $R_M(v) = \lambda_1$, where λ_1 is the minimal eigenvalue of M and $v = v_1$ is the corresponding eigenvector (and similarly for the maximum).

One can generalize this quotient to functionals, similar as in [3], by

$$R_J(u) := \frac{J(u)^2}{\|u\|^2}, \tag{10.6}$$

where $\| \cdot \|$ is the L^2 norm. We restrict the problem to the nondegenerate case where $J(u) > 0$, that is, u is not in the nullspace of J.

To find a minimizer, an alternative formulation can be used (as in the classical case):

$$\min_u \{J(u)\} \text{ s.t. } \|u\|^2 = 1. \tag{10.7}$$

Using Lagrange multipliers, the problem can be recast as

$$\min_u \{J(u) - \frac{\lambda}{2} \left(\|u\|^2 - 1 \right)\},$$

with the optimality condition

$$0 \in \partial J(u) - \lambda u,$$

which coincides with the eigenvalue problem (4.3). Note that with the additional constraint of u being in the orthogonal complement of the nullspace of J in the case of J being absolutely one-homogeneous, in which case one obtains the minimal nonzero eigenvalue as a minimum of the generalized Rayleigh quotient (10.7).

We now proceed to some theoretical results of the nonlinear case. The work by Benning and Burger in [3] considers more general variational reconstruction problems involving a linear operator in the data fidelity term, i.e.,

$$\min_u \frac{1}{2} \|Au - f\|_2^2 + t\, J(u),$$

and generalizes Eq. (4.3) to

$$\lambda A^* A u \in \partial J(u), \qquad \|Au\|_2 = 1,$$

in which case u is called a *singular vector*. Particular emphasis is put on the *ground states*

$$u^0 = \arg \min_{\substack{u \in \mathrm{kern}(J)^\perp, \\ \|Au\|_2 = 1}} J(u)$$

for semi-norms J, which were proven to be singular vectors with the smallest possible singular value. Although the existence of a ground state (and hence the existence of singular vectors) is guaranteed for all reasonable J in regularization methods, it was shown that the Rayleigh principle for higher singular values fails. This means that one does not necessarily have a decomposition of the signal into *orthogonal* nonlinear eigenfunctions, as in the linear case. This was shown earlier, where we found a spectral decomposition into eigenfunctions p (the subgradients along the flow), which were not orthogonal. This is the importance of the ϕ spectral component, which admits orthogonality, and has no direct analog to the linear case (it is composed of two eigenfunctions, but is not necessarily an eigenfunction).

In the setting of nonlinear eigenfunctions for one-homogeneous functionals, the Rayleigh principle for the second eigenvalue (given the smallest eigenvalue λ_1 and ground state u_1) leads to the optimization problem

$$\min_{u}\{J(u)\} \text{ s.t. } \|u\|^2 = 1, \langle u, u_1 \rangle = 0. \tag{10.8}$$

With appropriate Lagrange multipliers λ and μ we obtain the solution as a minimizer of

$$\min_{u}\{J(u) - \frac{\lambda}{2}\left(\|u\|^2 - 1\right) + \mu\langle u, u_1 \rangle\},$$

with the optimality condition

$$\lambda u - \mu u_1 = p \in \partial J(u).$$

We observe that we can only guarantee u to be an eigenvector of J if $\mu = 0$, which is not guaranteed in general. A scalar product with u_1 and the orthogonality constraint yields $\mu = -\langle p, u_1 \rangle$, which only needs to vanish if $J(u) = \|u\|$, but not for general one-homogeneous J.

An example of a one-homogeneous functional failing to produce the Rayleigh principle for higher eigenvalues (which can be derived from the results in [3]) is given by $J : \mathbb{R}^2 \to \mathbb{R}$,

$$J(u) = \|Du\|_1, \quad D = \begin{pmatrix} 1 & -2\varepsilon \\ 0 & \frac{1}{\varepsilon} \end{pmatrix},$$

for $0 < \varepsilon < \frac{1}{2}$. The ground state $u = (u^1, u^2)$ minimizes $|u^1 - 2\varepsilon u^2| + \frac{1}{\varepsilon}|u^2|$ subject to $\|u\|_2 = 1$. It is easy to see that $u = \pm(1, 0)$ is the unique ground state, since by the normalization and the triangle inequality

$$1 = \|u\|_2 \leq \|u\|_1 \leq |u^1 - 2\varepsilon u^2| + (1 + 2\varepsilon)|u^2|$$
$$\leq |u^1 - 2\varepsilon u^2| + \frac{1}{\varepsilon}|u^2|,$$

and the last inequality is sharp if and only if $u^2 = 0$. Hence, the only candidate v being normalized and orthogonal to u is given by $v = \pm(0, 1)$. Without restriction of generality we consider $v = (0, 1)$. Now, $Dv = \left(-2\varepsilon, \frac{1}{\varepsilon}\right)$ and hence

$$\partial J(v) = \{D^T(-1, 1)\} = \left\{(-1, 2\varepsilon + \frac{1}{\varepsilon})\right\},$$

which implies that there cannot be a $\lambda > 0$ with $\lambda v \in \partial J(v)$. A detailed characterization of functionals allowing for the Rayleigh principle for higher eigenvalues is still an open problem as well as the question whether there exists an orthogonal basis of eigenvectors in general. However, in the above example, we are not able to construct an orthogonal basis of eigenvectors — this will become more apparent from the characterization of eigenvectors in the next section.

10.4 Sparse Representation by Eigenfunctions

With our previous definitions, a convex functional induces a dictionary by its eigenfunctions. Let us formalize this notion by the following definition.

Definition 10.1 (*Eigenfunction Dictionary*) Let $\mathscr{D}_J$ be a dictionary of functions in a Banach space $\mathscr{X}$, with respect to a convex functional J, defined as the set of all eigenfunctions,

$$\mathscr{D}_J := \bigcup_{\lambda \in \mathbb{R}} \{u \in \mathscr{X} \mid \lambda u \in \partial J(u), \ \|u\|_2 = 1\}.$$

There are natural questions to be asked such as

1. What functions can be reconstructed by a linear combination of functions in the dictionary $\mathscr{D}_J$?
2. Is $\mathscr{D}_J$ an overcomplete dictionary?
3. Does $\mathscr{D}_J$ contain a complete orthonormal system?
4. How many elements are needed to express some type of functions (or how sparse are they in this dictionary)?

We notice that for many functionals, in particular semi-norms, there is a natural ambiguity between u and $-u$, which are both elements of $\mathscr{D}_J$.

So far, the above questions have not been investigated systematically and there is no general answer. However, from the existing examples, we expect the eigenfunctions of a convex one-homogeneous function to contain a basis, and often be extremely overcomplete. Regarding the third aspect, it would be desirable to have a criterion for when input data consisting of a linear combination of eigenfunctions can actually be decomposed into its original atoms.

For some simple cases, we can immediately give an answer to the questions raised above. First, note that in the case of $J(u) = \|u\|_1$ any $u = \frac{g}{\|g\|_2}$ with $g_i \in \{-1, 0, +1\}$ is an eigenfunction (except $g \equiv 0$). Thus, even after removing the sign ambiguity there still exist $3^n/2 - 1$ different eigenfunctions in $\mathbb{R}^n$. Remarkably, the number of eigenfunctions grows exponentially with the dimension of the space.

10.4.1 Total Variation Dictionaries

In the continuous setting, the result of Proposition 10.1 stating that any Haar wavelet function is a TV eigenfunction immediate yields the existence of a basis of eigenfunction in this case, too (see, for example, [4] for details on the wavelet reconstruction properties). Moreover, the explicit construction of eigenfunctions according to Proposition 4.1 along with the freedom to choose the $x_i \in \mathbb{R}$ arbitrarily yields that there are uncountably many eigenfunctions and the dictionary is overcomplete.

Without having the immediate connection to wavelets, the above results extend to the c-dimensional TV regularization:

Corollary 10.1 *For $J_{TV} : BV(\mathbb{R}^c) \to \mathbb{R}$ being the TV functional, any $f \in L^2(\mathbb{R}^c)$ can be approximated up to a desired error ε by a finite linear (respectively, conical) combination of elements from $\mathcal{D}_{J_{TV}}$. Moreover, $\mathcal{D}_{J_{TV}}$ is an overcomplete dictionary (so the linear combination is not unique).*

Proof First of all, we can approximate any L^2-function to arbitrary precision with a linear combination of a finite number of piecewise constant functions. Since the Borel sets are generated by balls, we can further construct an approximation to arbitrary precision with functions piecewise constant on balls, i.e., linear combinations of characteristic functions of balls. Since the latter are eigenfunctions of the total variation, i.e., elements of $\mathcal{D}_J$, the dictionary is complete. Moreover, since there are Cheeger sets not equal to a ball, and their characteristic functions are in $\mathcal{D}_J$, too, the dictionary is overcomplete.

While the properties of eigenfunctions with respect to the total variation regularization are fairly well understood, the question for the properties of a dictionary of eigenfunctions for general one-homogeneous regularization functionals remains open. In [5] we showed that the regularization $J(u) = \|Du\|_1$ for DD^* being diagonally dominant always yields the existence of a basis of eigenfunctions. We do, however, expect to have a significantly overdetermined dictionary of eigenfunctions. Unfortunately, classical ways of determining eigenfunctions – such as the Rayleigh principle – fail in the setting of generalized eigenfunctions (4.3) for non-quadratic J as we have seen in the previous section. Nevertheless, for absolutely one-homogeneous functions we can describe the set of eigenvectors more precisely.

10.4.2 Dictionaries from One-Homogeneous Functionals

In the following let us consider one-homogeneous functionals $J : \mathbb{R}^n \to \mathbb{R}$, which by duality can be represented as

$$J(u) = \sup_{q \in K} \langle q, u \rangle \tag{10.9}$$

for some convex set K. In order to avoid technicalities (related to a possible nullspace of J) we assume that K has nonempty interior. The following result provides a characterization of eigenvectors for nonzero eigenvalues:

Lemma 10.1 *Let J be defined by* (10.9) *and $p \in \partial K$ satisfy*

$$\langle p, p - q \rangle \geq 0 \quad \forall q \in K. \tag{10.10}$$

Then $u = \frac{p}{\|p\|}$ is an eigenvector with eigenvalue $\lambda = \|p\|$. Vice versa, if u is an eigenvector with eigenvalue $\lambda \neq 0$, then $p = \lambda u$ satisfies (10.10).

Proof We have $p \in \partial J(u)$ if and only if p is the maximal element in $\langle q, u \rangle$ over all $q \in K$. Hence, u is an eigenvector, i.e., a positive multiple of p if and only if (10.10) holds.

This result has a straightforward geometric interpretation: if $u = \lambda p$ is an eigenvector corresponding to $\lambda \neq 0$, then a ball centered at zero is tangential to ∂K at p. In other words, the hyperplane through p and orthogonal to p is tangential to ∂K, more precisely K lies on one side of this hyperplane.

To conclude this chapter, it appears there are many analogs and similarities between the linear and nonlinear case, and certain concepts are generalized in a very natural manner, such as a generalized Rayleigh quotient. However, there are some new concepts which seem to degenerate in the linear case and exist only in the nonlinear setting. Thus, eigenfunctions are not necessarily a basis and are not orthogonal in general, and the decomposition into orthogonal elements ϕ is required. Surely, far more investigations are needed to understand the new concepts and find the relations between them.

References

1. M. Burger, L. Eckardt, G. Gilboa, M. Moeller, Spectral Representations of One-Homogeneous Functionals, *Scale Space and Variational Methods in Computer Vision* (Springer, Berlin, 2015), pp. 16–27
2. G. Steidl, J. Weickert, T. Brox, P. Mrzek, M. Welk, On the equivalence of soft wavelet shrinkage, total variation diffusion, total variation regularization, and SIDEs. SIAM J. Numer. Anal. **42**(2), 686–713 (2004)
3. M. Benning, M. Burger, Ground states and singular vectors of convex variational regularization methods. Methods Appl. Anal. **20**(4), 295–334 (2013)
4. C.K. Chui, *An Introduction to Wavelets*, vol. 1 (Academic press, Cambridge, 1992)
5. Martin Burger, Guy Gilboa, Michael Moeller, Lina Eckardt, Daniel Cremers, Spectral decompositions using one-homogeneous functionals. SIAM J. Imaging Sci. **9**(3), 1374–1408 (2016)

Chapter 11
Future Directions

11.1 Spectral Total Variation Local Time Signatures for Image Manipulation and Fusion

We suggest a new approach to exploit comprehensive spectral TV-derived information to isolate and differentiate objects within an image, and apply it to image manipulation and fusion.

Given an image, we extract spectral TV local time signatures, based on the full $\phi(t, x)$ data. To handle this high-dimensional, redundant data, we reduce its dimensionality using clustering. This allows to partition the image into different regions or objects with common spectral TV features, to acquire an image map of semantically significant, distinct groups of objects.

Before clustering, we may denoise the image, enhance the strength and sparsity of time signatures, or choose certain image regions we wish to address. We then manipulate these regions using image matting or morphological operations.

Our framework stems from the edge-preserving, sparse spectral TV transform, and demonstrates the following useful attributes: locality, rotation and translation invariance, and responsiveness to size and *local* contrast.

Thus, we design an algorithm for isolating and differentiating objects within an image. These objects may be of different contrasts, sizes and structures, as well as super-positioned and multi-scaled objects. This is applicable for real images of complex nature and different modalities.

We, therefore, suggest a single unified framework with two major applications for image processing. For image fusion (Fig. 11.1), we construct saliency maps of highly contrasted objects from thermal or medical (MRI-T2) images, to be fused with the corresponding visible or different-modality medical image, respectively. For image manipulation of natural images with repetitive structures (Fig. 11.2), we can substitute various pro-processing techniques, such as image segmentation, edge detection or size differentiation, by extracting interesting distinct image structures, sharing common features. Our contribution lies in suggesting a single unified frame-

© Springer International Publishing AG, part of Springer Nature 2018
G. Gilboa, *Nonlinear Eigenproblems in Image Processing and Computer Vision*, Advances in Computer Vision and Pattern Recognition,
https://doi.org/10.1007/978-3-319-75847-3_11

Fig. 11.1 Thermal and visible image fusion example using our method. Upper left: thermal input image. Lower left: corresponding visible input image. Upper right: saliency map of semantically meaningful thermal objects extracted using spectral TV local time signatures. Lower right: fusion result of saliency map and corresponding visible image

Fig. 11.2 Example of manipulating a natural image with repetitive structures. Left: leaf input image. Middle: map of semantically significant structures extracted using spectral TV local time signatures. The extraction of leaf veins is allowed due to the properties of rotation and translation invariance and local contrast responsiveness. Right: leaf veins manipulated according to structures map

work, suitable for various useful applications, while relying merely on a few simple cornerstones.

11.2 Spectral AATV (Adapted Anisotropic Total Variation) and the AATV Eigenfunction Analysis

The spectral AATV approach is an extension of the spectral TV. While spectral TV is doing a great job in building a spectrum of texture "frequencies". The spectral AATV extends the spectrum to both texture "frequencies" and objects "frequencies".

The way it is done is by introducing the AATV functional to the spectral framework. The AATV functional itself is, for smooth functions

$$J_{AATV}(u) = \int_{\Omega} |\nabla_A u(x)| dx,$$

or generally

$$J_{AATV}(u) = \sup_{\xi^A \in C_c^{\infty}, \|\xi^A\|_{\infty} \leq 1} \int_{\Omega} u(x) \operatorname{div}_A \xi^A(x) dx.$$

where the adapted gradient and divergence defined by

$$\operatorname{grad}_A = \nabla_A \triangleq A\nabla, \quad \operatorname{div}_A = \nabla_A^T \triangleq \nabla^T A.$$

Here, $A(x) \in \mathbb{R}^{2 \times 2}$ is a symmetric matrix, spatially adapted, $A \succ 0$. A is constructed at each point based on the analysis of the structure tensor. At each location its first eigenvector is in the gradient direction and its second is in the level set direction, similar to anisotropic diffusion methods. The level-set direction is with eigenvalue 1 and the gradient direction is set with a smaller eigenvalue $0 < a < 1$. Thus the coordinated are transformed and rotated locally according to the local dominant direction.

Basically, the spectral framework relies on a flow of a general one-homogeneous functional J. The reason why the TV functional can not be used for the extraction of the objects "frequencies", is that along the flow, the TV changes the shapes within the image to shapes that have more and more (as the flow goes) of the functional eigenfunctions properties, which are convexedness and roundness.

The AATV functional transcend that problem by using the adapted matrix A to determine where to perform regular TV regularization (on constant areas), and where to perform high regularization (on edges); as it shown in Fig. 11.3.

The research field includes defining those new eigenfunctions and eigenvalues of the AATV functional while providing mathematical proofs and a new theory on this exciting new field.

The applications of the spectral AATV ranges from artistic manipulation of texture and objects DC level (Fig. 11.4), naive clustering, denoising, and with some ingenuity, to a reconstruction-based spectral AATV (Fig. 11.5).

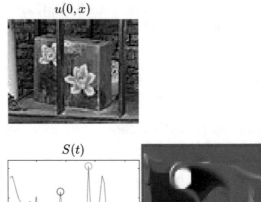

$$u(0, x)$$

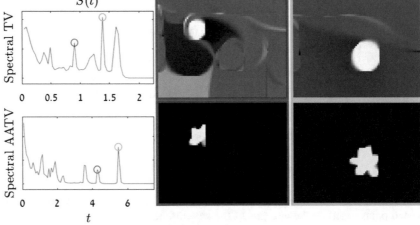

Spectral TV

$$S(t)$$

Spectral AATV

$$t$$

Fig. 11.3 AATV/TV spectrum comparison of a FlowerBox. Here we can see that the flowers, in both schemes, kept their location, but only the spectral AATV kept their exact boundaries

Fig. 11.4 Flowers DC level manipulation in a FlowerBox. In this example, we took the frequencies of the two flowers, and increased them from zero (top-left) by 33% from top to down, and left to right

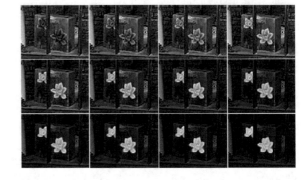

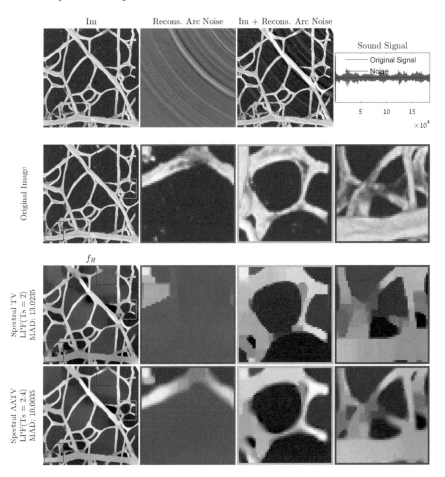

Fig. 11.5 Optoacoustic spectral AATV-based reconstruction. The output of the optoacoustic imaging device is a sound signal, arriving from numerous sensors. In this example, we added an arc noise by multiplying the received projection of one of the sensors by 5. One can see that when applying a strong LPF (which means strong regularization by the appropriate functional). The spectral AATV gives much better results by keeping much more details of this artery example

11.3 TV Spectral Hashing

Weiss et al. [1] suggested that a good code should satisfy three conditions: (a) It should be easily computed for a novel input (b) It should require a small number of bits to code the full dataset and (c) It would map similar items to similar binary codewords. To achieve this, the following optimization problem was formulated. Let $x^1, \ldots, x^n$ given vectors, k desired bit rate and a similarity $W_{ij} = exp\left(-\left\|x^i - x^j\right\|^2 / \varepsilon\right)$, the binary codewords are given by

$$\text{argmin} \sum_{ij} w_{ij} \left\| v^i - v^j \right\|_2^2, \tag{11.1}$$

$$s.t: \ v^1, \ldots, v^n \in \{-1, 1\}^k, \tag{11.2}$$

$$\sum_i y^i = 0, \tag{11.3}$$

$$\frac{1}{n} \sum_i v^i \left(v^i\right)^T = I. \tag{11.4}$$

To solve this NP-hard problem it was suggested to first relax it by excluding the constraint $v^i \in \{-1, 1\}^k$. The solution to the relaxed problem is given by the first k eigenvectors of the 2-Laplacian. To approximate the solution of (the original) problem (11.1), the final solution is derived by thresholding these eigenfunctions. Unfortunately, following the thresholding operation the constraints do not hold anymore, moreover, the corresponding objective function in most cases does not attain its minimum. To overcome these problems, we propose an alternative formulation and solution. We reformulate Eqs. (11.1)–(11.4) as follows:

$$\text{argmin} \sum_{i=1}^k F_1\left(u^i\right) \tag{11.5}$$

$$s.t: \ u^1, \ldots, u^k \in \{-1, 1\}^n, \tag{11.6}$$

$$median\left(u^i\right) = 0, \tag{11.7}$$

$$\frac{1}{n} \left\langle u^i, u^j \right\rangle = \delta_{ij}. \tag{11.8}$$

Proposition 11.1 *The two formulations are equivalent.*

Proof First, let us look at the two sets of solutions in a matrix form by

$$\begin{pmatrix} v_1^1 & v_2^1 & \cdots & v_k^1 \\ v_1^2 & v_2^2 & \cdots & v_k^2 \\ \vdots & \vdots & \ddots & \vdots \\ v_1^n & v_2^n & \cdots & v_k^n \end{pmatrix} = \begin{pmatrix} u_1^1 & u_1^2 & \cdots & u_1^k \\ u_2^1 & u_2^2 & \cdots & u_2^k \\ \vdots & \vdots & \ddots & \vdots \\ u_n^1 & u_n^2 & \cdots & u_n^k \end{pmatrix}, \tag{11.9}$$

where v^i and u^j represent the ith row and jth column of the same $n \times k$ matrix. We can now immediately see the dimension agreement of the two binary solutions which relates to the first constraint. The second constraint can be simply represented by taking the sum of each column to be zero, where the binary constraint, in this case, forces a median zero of each column (we assume here an even n and take the mean of the two values closest to the median). The third constraint can be simply seen by the following equalities:

$$\left(\sum_{i=1}^{n} v^i \left(v^i\right)^T \right)_{ml} = \sum_{i=1}^{n} v_m^i v_l^i = \sum_{i=1}^{n} u_i^m u_i^l = \langle u^m, u^l \rangle. \qquad (11.10)$$

The equivalence of the objective minimizers can be seen by using the fact that for binary vectors of size n, $v^i \in \{-1, 1\}^n$, we have $(v_{il} - v_{jm}) \in \{-2, 0, 2\}, \forall i, j, l, m$. This yields the following identities:

$$\left\| v^i - v^j \right\|_2^2 = 2 \left\| v^i - y^j \right\|_1 = 2 \sum_{l=1}^{k} \left| v_l^i - v_l^j \right| = 2 \sum_{l=1}^{k} \left| u_i^l - u_j^l \right|, \quad \left\| u^l \right\|_1 = n. \qquad (11.11)$$

We can then immediately derive that

$$\operatorname{argmin} \sum_{ij} w_{ij} \left\| v_i - v_j \right\|_2^2 = \operatorname{argmin} \sum_{ij} w_{ij} \sum_{l=1}^{k} \frac{\left| u_i^l - u_j^l \right|}{\left\| u^l \right\|_1} = \operatorname{argmin} \sum_{l=1}^{k} \frac{J_{TV} \left(u^l\right)}{\left\| u^l \right\|_1}. \qquad (11.12)$$

Theorem 11.1 *If $u_1, \ldots, u_n$ are nonzero everywhere 1-eigenfunctions associated with unique 1-eigenvalues $\lambda_1, \ldots, \lambda_n$, then $sgn(u_1), \ldots, sgn(u_n)$ are local solution to the following problem:*

$$\operatorname{argmin} \sum_{i=1}^{k} F_1 \left(u^i\right), \qquad (11.13)$$

$$s.t.: \ u^1, \ldots, u^k \in \{-1, 1\}^n, \qquad (11.14)$$

$$median \left(u^i\right) = 0, \qquad (11.15)$$

$$\frac{1}{n} \langle u^i, u^j \rangle \approx \delta_{ij}. \qquad (11.16)$$

A proof will be published in a future publication with Tal Feld.

11.4 Some Open Problems

This field is in its early research stages and there are many theoretical questions, as well as possibilities for new applications and uses of the theory in new fields. We outline a few theoretical and applied open problems:

1. What are the spectral components ϕ? Is there a general way to formalize them for any convex or even nonconvex functional?
2. How far can this theory go in the nonconvex case? What are the conditions to have only one family of eigenfunctions with a prototypical decay?
3. Regarding orthogonality of ϕ, the current proof is only within a restricted setting and in finite dimensions. Can it be generalized to infinite dimensions? to any spatial dimension? to graphs?
4. With respect to applications, what are the relations of multiple 1-Laplacian eigenfunctions to the Cheeger cut problem? How can this be used to solve better segmentation and clustering problems?
5. What one-homogeneous functionals, other than total variation, can be used in a spectral manner for various applications?
6. Can additional linear concepts be generalized in the nonlinear setting? For instance, is there an analog of Fourier duality with respect to convolution and multiplication (in the spatial and transform domains)?

Reference

1. Y. Weiss, A. Torralba, R. Fergus. Spectral hashing. In *Advances in neural information processing systems*. (2009)

Appendix A
Numerical Schemes

All's well that ends well

A.1 Derivative Operators

We will use the 1D notations (generalization to any dimension is straightforward). Those are the common first-order derivative operators:

Forward difference: $D^+ u_i = \frac{u_{i+1} - u_i}{h}$
Backward difference: $D^- u_i = \frac{u_i - u_{i-1}}{h}$
Central difference: $D^0 u_i = \frac{u_{i+1} - u_{i-1}}{2h}$.

Remark For smooth functions, the central difference has a second-order accuracy (proportional to h^2), whereas the forward and backward operators are first order (proportional to h). However, as images are not smooth and the fact that derivatives of high oscillations are not estimated well in the central scheme, the forward or backward schemes are usually used for our PDE's.

Time: Can be discretized similarly. Usually a backward difference is used (since we cannot predict the future): $D_t^+ u_i^n = \frac{u_i^n - u_i^{n-1}}{\Delta t}$.

Partial derivative: As in the continuous case, the derivative is taken only with respect to one variable. For instance in the two dimensional case, for $u_{i,j}$ which is the discretized version of $u(x, y)$, the forward difference in the x direction is: $D_x^+ u_{i,j} = (u_{i+1,j} - u_{i,j})/h$.

Higher order derivatives: For higher order, one can use a composition of the first derivatives operators. For instance, the standard second derivative operator (central) is

$$D^2 u_i = D^+(D^- u_i) = \frac{u_{i+1} + u_{i-1} - 2u_i}{h^2}.$$

© Springer International Publishing AG, part of Springer Nature 2018
G. Gilboa, *Nonlinear Eigenproblems in Image Processing and Computer Vision*, Advances in Computer Vision and Pattern Recognition,
https://doi.org/10.1007/978-3-319-75847-3

This can also be seen as twice the central derivative at half-pixel resolution: $D_{1/2}^0 u_i = \frac{u_{i+1/2} - u_{i-1/2}}{h}$. And therefore $D^2 u_i = D_{1/2}^0 (D_{1/2}^0 u_i)$.

A.2   Discretization of PDE's

A.2.1   Discretized Differential Operators

There are many ways to discretize PDE's (e.g., finite elements, spectral methods). We will use the more simple *finite difference* schemes which are easy to implement, understand, and analyze.

A continuous function $u(t; x)$ is discretized to u_i^n, where $t = n\Delta t$ and $x = ih$, (n and i are usually integers). We assume to have a discrete regular grid with a step size h (in principle, a different h can be used for each dimension, we will assume the spatial discretization is the same in all directions). Δt is the time step.

Remark For images, we usually simply assume a unit step $h = 1$ between two adjacent pixels.

A.2.2   Evolutions

We use the backward difference for the time evolution: $u_t \approx \frac{u(t+\Delta t) - u(t)}{\Delta t}$. For an evolution of the form $u_t = F(u)$, the general iteration scheme is

$$u^{n+1} = u + \Delta t F(u).$$

There are several ways to model the flow, depending on the time taken for u on the right-hand side.

Explicit scheme: The right side depends solely on the previous iteration,

$$u_i^{n+1} = u_i^n + \Delta t F(u^n).$$

Very convenient to solve but requires small time steps (see CFL condition below).

Implicit scheme: $F()$ on the right side depends on the current iteration,

$$u_i^{n+1} = u_i^n + \Delta t F(u^{n+1}).$$

A system of linear equations are needed to solve this. Harder, good for large time steps.

Semi-implicit scheme: A combination of the two above schemes. For instance Weickert gives the following $\alpha-$ semi-implicit model (with $\alpha \in [0, 1]$) [1] p. 102:

$$u_i^{n+1} = u_i^n + \Delta t\, A(u^n)(\alpha u^{n+1} + (1 - \alpha)u^n),$$

where A depends nonlinearly on u. This enjoys the unconditional time step stability afforded by implicit schemes without the need to solve large systems of equations.

A.2.3   CFL Condition

For explicit diffusion schemes (linear and nonlinear) with 4-neighbor connectivity, the time step is bounded by

$$\Delta t \leq \frac{0.25 h^2}{\max_x c(x)}$$

This is a consequence of the CFL condition, which is a general criterion for evolution of PDE's, called after Courant, Friedrichs, and Lewy [2].

For example, in a standard implementation of Perona–Malik where $h = 1, c() \leq 1$ we have $\Delta t \leq 0.25$.

In the more general case, with M neighbors connectivity, p order PDE we have $\Delta t \leq \frac{h^p}{M \max_x c(x)}$.

A.3   Basic Numerics for Solving TV

Recent sophisticated numerical schemes to solve TV efficiently will be presented in the next section. We describe below three more classical methods: the simplest explicit method, Vogel–Oman lagged diffusivity [3] and Chambolle's projection [4]. We begin by defining discrete gradient and divergence operators in 2D (using the notations of [4]). We assume a $N \times N$ pixel grid with indices $i, j = 1, .. , N$ specifying the row and column.

Gradient: Let the discrete gradient for pixel i, j be defined as

$$(\nabla u)_{i,j} := \left((\nabla u)_{i,j}^1, (\nabla u)_{i,j}^2 \right), \tag{A.1}$$

where

$$
(\nabla u)_{i,j}^1 := \begin{cases} u_{i+1,j} - u_{i,j}, & \text{if } i < N, \\ 0, & \text{if } i = N, \end{cases}
$$
$$
(\nabla u)_{i,j}^2 := \begin{cases} u_{i,j+1} - u_{i,j}, & \text{if } j < N, \\ 0, & \text{if } j = N, \end{cases}
\tag{A.2}
$$

Divergence: Let a vector p be defined as $p = (p^1, p^2)$ (where p^k is a discrete 2D image). The discrete divergence is

$$(\text{div } p)_{i,j} = \begin{cases} p_{i,j}^1 - p_{i-1,j}^1, & \text{if } 1 < i < N, \\ p_{i,j}^1, & \text{if } i = 1, \\ -p_{i-1,j}^1, & \text{if } i = N, \end{cases} + \begin{cases} p_{i,j}^2 - p_{i,j-1}^2, & \text{if } 1 < j < N, \\ p_{i,j}^2, & \text{if } j = 1, \\ -p_{i,j-1}^2, & \text{if } j = N. \end{cases}$$

$$(\text{A.3})$$

This divergence definition admits an adjoint relation with the gradient defined above: $\langle -\text{div } p, u \rangle = \langle p, \nabla u \rangle$.

A.3.1 Explicit Method

Approximates $J_{TV-\varepsilon}(u) + \lambda \| f - u \|_{L^2}^2$. Let

$$F(u) = \text{div}\left(\frac{\nabla u}{\sqrt{|\nabla u|^2 + \varepsilon^2}}\right) + 2\lambda(f - u), \qquad (\text{A.4})$$

where ∇, and div are the discrete operators defined in Eqs. (A.1), and (A.3), respectively. Δt is bounded by the CFL condition.

Algorithm: initialize with $u^0 = f$, set $n = 0$. Evolve until convergence:

$$u^{n+1} = u^n + \Delta t F(u^n).$$

A.3.2 Lagged Diffusivity

Approximates $J_{TV-\varepsilon}(u) + \lambda \| f - u \|_{L^2}^2$.

We write the Euler–Lagrange equation:

$$\text{div}\left(\frac{\nabla u}{\sqrt{|\nabla u|^2 + \varepsilon^2}}\right) + 2\lambda(f - u) = 0.$$

We assume to have a previous value u^n (initializing with $u_0 = f$) and compute a new u^{n+1}. We can fix the diffusivity and define the following linear operator, which operates on v:

$$L_u v = \text{div}\left(\frac{\nabla}{\sqrt{|\nabla u|^2 + \varepsilon^2}}\right) v.$$

Using the E–L equation, changing order of variables, constructing L_{u^n} based on the previous iteration and dividing by 2λ we get

$$\left(1 - \frac{1}{2\lambda} L_{u^n}\right) u^{n+1} = f. \qquad (\text{A.5})$$

Thus, we have to solve a linear system of equations. The system is very sparse and there are several efficient ways to solve this. See more details in [3, 5].

A.3.3 Chambolle's Projection Algorithm

Chambolle proposed a fixed point method, based on a projection algorithm in [4] for solving the ROF problem described in [6]. A minimizer for the discrete ROF problem can be computed by the following iterations:

$$p_{i,j}^{n+1} = \frac{p_{i,j}^{n} + \tau(\nabla_{wd}(\text{div}\,(p^{n}) - 2\lambda f))_{i,j}}{1 + \tau|(\nabla(\text{div}\,(p^{n}) - 2\lambda f))_{i,j}|}, \tag{A.6}$$

where the initialization is $p^{0} = 0$ and $0 < \tau \leq 1/8$. The solution is $u = f - \frac{1}{2\lambda}\,\text{div}\,(p)$.

A.4 Modern Optimization Methods

Modern general optimization methods often involve the prox operator. We will briefly review them here. These methods and their variants can be used to solve most convex models which appear in image processing and computer vision.

A.4.1 The Proximal Operator

The proximal operator $\text{prox}_{f,\lambda} : \mathbb{R}^n \to \mathbb{R}^n$ is

$$\text{prox}_{f,\lambda}(v) = \text{argmin}_x \left\{ f(x) + \frac{1}{2\lambda}\|x - v\|^2 \right\}, \tag{A.7}$$

where $\|\cdot\|$ is the standard Euclidean (ℓ^2) norm and $\lambda > 0$. This operator yields a compromise between minimizing f and being close to v (the larger λ is, the closer it is to the minimizer of f). The Euler–Lagrange yields for $\lambda \to 0$ $\text{prox}_{f,\lambda}(v) \approx v - \lambda \nabla f$, therefore the operator can be interpreted as a step in a gradient descent.

A.4.1.1   Properties

Fixed points iterations of the proximal operator are minimizers of f. $\text{prox}_{f,\lambda}(x^*) = x^*$ if and only if x^* minimizes f.

If f is separable across two variables, $f(x) = \varphi(x) + \psi(y)$ then

$$\text{prox}_{f,\lambda}(v, w) = (\text{prox}_{\varphi,\lambda}(v), \text{prox}_{\psi,\lambda}(w)).$$

Similarly, in the general case if we can write f as $f(x) = \sum_{i=1}^{n} f_i(x_i)$, then

$$(\text{prox}_{f,\lambda}(v))_i = \text{prox}_{f_i,\lambda}(v_i).$$

A.4.2   Examples of Proximal Functions

In general if f is a norm and B is the unit ball of the dual norm we have

$$\text{prox}_{f,\lambda}(v) = v - \lambda \Pi_B(v/\lambda),$$

where Π_B is a projection onto the ball B.

Let $f = \| \cdot \|_2$ be the Euclidean norm in $\mathbb{R}^n$ (reminder, the dual of ℓ^2 is ℓ^2). A projection Π_B onto the unit ball B is

$$\Pi_B(v) = \begin{cases} v/\|v\|_2 & , \|v\|_2 > 1 \\ v & , \|v\|_2 \leq 1. \end{cases} \tag{A.8}$$

The proximal operator is thus

$$\text{prox}_{f,\lambda}(v) = (1 - \lambda/\|v\|_2)^+ v = \begin{cases} (1 - \lambda/\|v\|_2)v & , \|v\|_2 > \lambda \\ 0 & , \|v\|_2 \leq \lambda. \end{cases} \tag{A.9}$$

For the ℓ^1 norm, $f = \| \cdot \|_1$, we get elementwise soft thresholding. The proximal operator is

$$\text{prox}_{f,\lambda}(v)_i = \begin{cases} v_i - \lambda & , v_i > \lambda \\ 0 & , |v_i| \leq \lambda \\ v_i + \lambda & , v_i < -\lambda. \end{cases} \tag{A.10}$$

We now outline the three most common optimization methods used today for solving convex problems: ADMM (Split-Bregman [7]), FISTA [8] and Chambolle–Pock [9].

A.4.3 ADMM

Consider the problem of finding x which minimizes

$$f(x) + g(x) \tag{A.11}$$

where f, g are (closed, proper) convex functionals from $\mathbb{R}^n \to \mathbb{R}$, which are not necessarily smooth.

The alternating direction method of multipliers (ADMM) to solve this problem is as follows. Initialize with $x^0 = z^0 = u^0 = 0$. Set $k = 0$, $\lambda = \lambda_0$. Do the following iterations until convergence:

1. $x^{k+1} = \text{prox}_{f,\lambda}(z^k - u^k)$.
2. $z^{k+1} = \text{prox}_{g,\lambda}(x^{k+1} + u^k)$.
3. $u^{k+1} = u^k + x^{k+1} - z^{k+1}$.

Here, x^k and z^k converge to each other and to optimality.

A variant of the ADMM is very popular in denoising total variation and similar functionals, referred to as the *Split-Bregman* algorithm by Goldstein and Osher [7], see a review on this and the relation to ADMM in [10].

A.4.4 FISTA

In [8], the following algorithm was proposed. Initialize with $x^0 = y^0$, $t^1 = 1$. Set $k = 0$, L is the Lipschitz constant of ∇f. Do the following iterations until convergence:

1. $x^k = \text{argmin}_x \left\{ g(x) + \frac{L}{2} \| x - (y - \frac{1}{L} \nabla f(y)) \|^2 \right\}$.
2. $t^{k+1} = (1 + \sqrt{1 + 4t_k^2})/2$.
3. $y^{k+1} = x^k + \frac{t_k - 1}{t_{k+1}}(x_k - x^{k-1})$.

A.4.5 Chambolle–Pock

Given K a linear operator (such as a convolution blur kernel) the algorithm aims at minimizing

$$f(Kx) + g(x) \tag{A.12}$$

This is done by alternating between the primal and dual spaces. In the dual space, the problem is to find y which *maximizes* the dual expression

$$- f^*(y) - g^*(-K^T y). \tag{A.13}$$

Initialize with $x^0 = y^0 = \bar{x}^0 = 0$, $\tau = \sigma = \theta = 1$. Set $k = 0$. Do the following iterations until convergence:

1. $y^{k+1} = \text{prox}_{f^*,\sigma}(y^k + \sigma K \bar{x}^k)$.
2. $x^{k+1} = \text{prox}_{g,\tau}(x^k - \tau K^T y^{k+1})$.
3. $\bar{x}^{k+1} = x^{k+1} + \theta(x^{k+1} - x^k)$.

Faster convergence are attained in accelerated versions (which are valid in a subset of the cases) where *theta* and the other parameters change every iteration:

$$.\theta^k = 1/\sqrt{1 + 2\gamma\tau^k}, \quad \tau^{k+1} = \theta^k\tau^k, \quad \sigma^{k+1} = \sigma^k/\theta^k,$$

(with some $\gamma > 0$.).

The algorithm initiated by Pock–Cremers–Bischof–Chambolle for the Mumford–Shah functional [11]. Later it was analyzed and generalized in [9].

A.5 Nonlocal Models

A.5.1 Basic Discretization

Let u_i denote the value of a pixel i in the image ($1 \leq i \leq N$), $w_{i,j}$ is the sparsely discrete version of $w(x, y)$. We use the neighbors set notation $j \in \mathcal{N}_i$ defined as $j \in \mathcal{N}_i := \{j : w_{i,j} > 0\}$.

Let ∇_{wd} be the discretization of ∇_w:

$$\nabla_{wd}(u_i) := (u_j - u_i)\sqrt{w_{i,j}}, \quad j \in \mathcal{N}_i \tag{A.14}$$

Let div_{wd} be the discretization of div_w:

$$\text{div}_{wd}(p_{i,j}) := \sum_{j \in \mathcal{N}_i}(p_{i,j} - p_{j,i})\sqrt{w_{i,j}}. \tag{A.15}$$

The discrete inner product for functions is $< u, v >:= \sum_i(u_i v_i)$ and for vectors we have the discretized dot product $(p \cdot q)_i := \sum_j(p_{i,j}q_{i,j})$ and inner product $< p, q >:= \sum_i \sum_j(p_{i,j}q_{i,j})$. The vector magnitude is therefore $|p|_i := \sqrt{\sum_j(p_{i,j})^2}$.

Weights Discretization:

The weights are discretized as follows: we take a patch around a pixel i, compute the distances $(d_a)_{i,j}$, a discretization of $d_a(x, y)$, Eq. (2.21), to all the patches in the search window and select the k closest (with the lowest distance value). The number of neighbors k is an integer proportional to the area γ. For each selected neighbor j, we assign the value 1 to $w_{i,j}$ and to $w_{j,i}$. A maximum of up to $m = 2k$ neighbors for each pixel is allowed in our implementation. A reasonable setting is to take 5×5 pixel patches, a search window of size 21×21 and $m = 10$.

A.5.2 Steepest Descent

In this convex framework, one can resort as usual to a steepest descent method for computing the solutions. One initializes u at $t = 0$, e.g., with the input image: $u|_{t=0} = f$, and evolves numerically the flow:

$$u_t = -\partial_u J_d - \partial_u H_d(f, u),$$

where $\partial_u J_d$ is the discretized version of Eq. (2.35) or (2.36) and $H_d(f, u)$ is the discretized fidelity term functional. As in the local case, here also one should introduce a regularized version of the total variation: $\phi(s) = \sqrt{s + \epsilon^2}$ (where s is the square gradient magnitude). Thus, the E–L equations are well defined, also for a zero gradient.

A.5.2.1 Nonlocal ROF

Chambolle's projection algorithm [4] for solving ROF [6] can be extended to solve nonlocal ROF.

A minimizer for the discrete version of Eq. (2.41) can be computed by the following iterations (fixed point method):

$$p_{i,j}^{n+1} = \frac{p_{i,j}^n + \tau (\nabla_{wd}(\mathrm{div}_{wd}(p^n) - 2\lambda f))_{i,j}}{1 + \tau |(\nabla_{wd}(\mathrm{div}_{wd}(p^n) - 2\lambda f))_{i,j}|}, \tag{A.16}$$

where $p^0 = 0$, and the operators ∇_{wd} and div_{wd} are defined in (A.14) and (A.15), respectively. The solution is $u = f - \frac{1}{2\lambda} \mathrm{div}_{wd}(p)$.

Theorem *The algorithm converges to the global minimizer as $n \to \infty$ for any* $0 < \tau \le \frac{1}{\|\mathrm{div}_{wd}\|_{L^2}^2}$.

For the proof see [12].

A Bound on τ

The bound on τ depends on the operator norm $\|\mathrm{div}_{wd}\|^2$ which is a function of the weights $w_{i,j}$. As the weights are image dependent, so is $\|\mathrm{div}_{wd}\|^2$.

Let m be the maximal number of neighbors of a pixel, $m := \max_i \{\sum_j (\mathrm{sign}(w_{i,j}))\}$. If the weights are in the range $0 \le w_{i,j} \le 1 \ \forall i, j$, then for $0 < \tau \le \frac{1}{4m}$ the algorithm converges.

We need to show that $\| \text{div}_{wd} \|^2 \leq 4m$:

$$\| \text{div}_{wd}(p) \|^2 = \sum_i \left(\sum_j (p_{i,j} - p_{j,i}) \sqrt{w_{i,j}} \right)^2$$
$$\leq 2 \sum_i \left(\sum_j (p_{i,j}^2 + p_{j,i}^2) \right) \left(\sum_j w_{i,j} \right)$$
$$\leq 4 \max_i \left(\sum_j w_{i,j} \right) \sum_i \sum_j p_{i,j}^2$$
$$\leq 4m \| p \|^2.$$

A.5.2.2   Nonlocal TV-L1

To solve (2.42), we generalize the algorithm of [13]. We consider the problem:

$$\inf_{u,v} \left\{ J_{NL-TV}(u) + \frac{1}{2\alpha} \| f - u - v \|_{L^2}^2 + \lambda \| v \|_{L^1} \right\} \tag{A.17}$$

The parameter α is small so that we almost have $f = u + v$, thus (A.17) is a very good approximation of (2.42). We can solve the discretized version of (A.17) by iterating:

- v being fixed (we have a nonlocal ROF problem), find u using the nonlocal Chambolle's projection algorithm:

$$\inf_u \left(J_{NL-TV}(u) + \frac{1}{2\alpha} \| f - u - v \|_{L^2}^2 \right)$$

- u being fixed, find v which satisfies:

$$\inf_v \frac{1}{2\alpha} \| f - u - v \|_{L^2}^2 + \lambda \| v \|_{L^1}.$$

The solution for v is given by soft-thresholding $f - u$ with $\alpha\lambda$ as the threshold, denoted by $ST_{\alpha\lambda}(f - u)$, where

$$ST_\beta(q) := \begin{cases} q - \beta, \; q > \beta \\ 0, \quad\;\; |q| \leq \beta \\ q + \beta, \; q < -\beta. \end{cases} \tag{A.18}$$

The algorithm converges to the global minimizer as $n \to \infty$ for any $0 < \tau \leq \frac{1}{\| \text{div}_{wd} \|_{L^2}^2}$.

A similar extension to the iterative projection algorithm of [14] can be made also to the nonlocal setting.

Glossary

Ω Image domain.

$\partial\Omega$ Image boundary.

$f(x)$ or f Input image.

$u(x)$ or $u(x,t)$ or u Solution, evolved image.

$n(x)$ or n Noise.

g_σ Gaussian of standard deviation σ.

n Direction normal to the boundary.

$x = (x_1, .., x_N)$ Spatial coordinates (in N dimensions).

t Time (scale).

dt Time step (infinitesimal).

Δt Discrete time step.

∇ Gradient.

Δ Laplacian.

div Divergence.

$\langle \cdot, \cdot \rangle$ inner/dual product.

X_y or $\partial_y X$ or $\frac{\partial X}{\partial y}$ Partial derivative of X with respect to y.

X_{yy} or $\partial_{yy} X$ Partial second derivative of X with respect to y.

J or $J(u)$ Regularizing functional.

J_{TV} The total variation functional.

J_D The Dirichlet functional.

∂J Subdifferential (possibly a multivalued set).

p or $p(u)$ or $p \in \partial J(u)$ Subgradient (an element in the subdifferential).

λ Eigenvalue.

G. Gilboa, *Nonlinear Eigenproblems in Image Processing and Computer Vision*, Advances in Computer Vision and Pattern Recognition, https://doi.org/10.1007/978-3-319-75847-3

T or $T(u)$ A general nonlinear operator.

L or $L(u)$ A linear operator.

$\mathcal{N}(J)$ Nullspace of J.

ϕ or $\phi(t)$ or $\phi(t, x)$ Spectral response.

S or $S(t)$ Spectrum.

ω Frequency.

$\mathcal{F}$ Fourier transform.

$*$ Convolution.

References

1. J. Weickert, *Anisotropic Diffusion in Image Processing* (Teubner-Verlag, Stuttgart, 1998)
2. Richard Courant, Kurt Friedrichs, Hans Lewy, Über die partiellen differenzengleichungen der mathematischen physik. Mathematische Annalen **100**(1), 32–74 (1928)
3. R.V. Vogel, M.E. Oman, Iterative methods for total variation denoising. SIAM J. Sci. Comput. **17**(1), 227–238 (1996)
4. A. Chambolle, An algorithm for total variation minimization and applications. JMIV **20**, 89–97 (2004)
5. T.F. Chan, P. Mulet, On the convergence of the lagged diffusivity fixed point method in total variation image restoration. SIAM J. Numer. Anal. **36**(2), 354–367 (1999)
6. L. Rudin, S. Osher, E. Fatemi, Nonlinear total variation based noise removal algorithms. Physica D **60**, 259–268 (1992)
7. T. Goldstein, S. Osher, The split bregman method for l1-regularized problems. SIAM J. Imaging Sci. **2**(2), 323–343 (2009)
8. Amir Beck, Marc Teboulle, A fast iterative shrinkage-thresholding algorithm for linear inverse problems. SIAM J. Imaging Sci. **2**(1), 183–202 (2009)
9. A. Chambolle, T. Pock, A first-order primal-dual algorithm for convex problems with applications to imaging. J. Math. Imaging Vis. **40**(1), 120–145 (2011)
10. Ernie Esser, Applications of lagrangian-based alternating direction methods and connections to split bregman. CAM Rep. **9**, 31 (2009)
11. T. Pock, D. Cremers, H. Bischof, A. Chambolle, An algorithm for minimizing the mumford-shah functional, in *2009 IEEE 12th International Conference on Computer Vision* (IEEE, 2009), pp. 1133–1140
12. G. Gilboa, S. Osher, Nonlocal operators with applications to image processing. SIAM Multi-scale Model. Simul. **7**(3), 1005–1028 (2008)
13. J.F. Aujol, G. Gilboa, T. Chan, S. Osher, Structure-texture image decomposition - modeling, algorithms, and parameter selection. Int. J. Comput. Vis. **67**(1), 111–136 (2006)
14. J.F. Aujol, G. Aubert, L. Blanc-Féraud, A. Chambolle, Image decomposition into a bounded variation component and an oscillating component. JMIV **22**(1) (2005)

Index

© Springer International Publishing AG, part of Springer Nature 2018

G. Gilboa, *Nonlinear Eigenproblems in Image Processing and Computer Vision*, Advances in Computer Vision and Pattern Recognition, https://doi.org/10.1007/978-3-319-75847-3

Printed in the United States
By Bookmasters